G. H. Perkins

Insects injurious to the American elm

G. H. Perkins

Insects injurious to the American elm

ISBN/EAN: 9783743323261

Manufactured in Europe, USA, Canada, Australia, Japa

Cover: Foto ©berggeist007 / pixelio.de

Manufactured and distributed by brebook publishing software
(www.brebook.com)

G. H. Perkins

Insects injurious to the American elm

From Eleventh Report of Vermont State Board of Agriculture.

Insects Injurious

: : : TO THE : : :

AMERICAN ELM.

G. H. PERKINS, Ph. D.

MONTPELIER, VT.
ARGUS AND PATRIOT JOB PRINTING HOUSE.
1890.

INJURIOUS INSECTS.

G. H. Perkins, Ph, D.

Professor of Natural History in the University of Vermont. Entomologist to the State Agricultural Experiment Station.

METHODS OF PREVENTING OR CHECKING THE ATTACKS OF INJURIOUS INSECTS.—I. INSECT ENEMIES OF INJURIOUS INSECTS.

In order that we may meet the attacks of destructive insects with any chance of success, it is necessary, first, that we understand something of the structure, habits, and mode of development of each species ; and, second, that we be able to devise some method by which the enemy may be driven off, or, if possible, exterminated. Usually the war must be one of extermination, for, as every farmer and only too well knows, insects when driven off very often return speedily, and with re-enforcement. Thus the study of insects or of insecticides must be carried on together. New insects are every now and then appearing, some of them injurious, some beneficial, some neither. Most injurious insects are attacked in one way or another by other insects. These may be either parasitic upon eggs, larva, pupa or imago, that is, upon any of the different stages which ordinarily occur in the life of insects, and sometimes an injurious insect is attacked in each of these stages by a different enemy. The sudden decrease in the number of insects, which every one has now and then observed, is often due to the destruction caused by parasitic insects. I remember a season some years ago, when canker worms were most ruinously abundant, so much so, that the citizens of the infested town were quite in despair concerning elms and other trees, and the prospects for the next season were certainly discouraging ; but, to the agreeable surprise of every one, the next season brought not only no increase of the canker worm, but so very marked a decrease as to well-nigh render efforts for the prevention of further damage to the trees unnecessary, and, for a number of years after this, there was no return of the pest. This deliverance was due to the attacks of certain parasites, which so effectually destroyed the canker worms as to nearly drive

4

them from the locality. So complete a removal of the evil does not, indeed, often occur, but most of our injurious insects are kept more or less in check by parasites. Some insects are attacked by three or four, or, it may be, ten or twelve parasites. It is obvious that an investigation of this subject of parasitism is of the highest importance. It is, of course, a blind and bungling method of warfare to fight against all insects indiscriminately. We must always remember that there are insects which are of value to the farmer, and the destruction of these is as great a mistake as it is to allow an injurious insect to escape. It is a most fortunate law which seems to prevail throughout the insect world, that vegetable-feeding insects, which are mostly injurious, are almost always attacked by one or more, sometimes many, parasites, and as the pernicious species increase, the parasite is very likely to increase, and often the latter destroys the whole race of the former,

It is not quite easy, without occupying too great space, to give definite descriptions of these parasitic insects which are our friends ; but the subject is one of so much importance that, with the aid of the accompanying illustrations, an attempt will be made to give a general idea of some of the more common beneficial insects, although it will not be possible to give anything like a complete account of them, or to even mention more than a few. However, the principal classes are illustrated. Not only parasitic, but predatory insects, those which are the lions and tigers of the insect world, going about seeking the bugs they would devour, must be noticed, and it is sincerely hoped that what information is herein given may be of some assistance in enabling the farmer to distinguish his friends from his foes. The ichneumons, bristle tails, etc., mentioned and figured must serve as types of the groups to which they belong, and, although not strictly correct, yet it will on the whole be safe to consider all similar insects as friends ; for, while some of the gall-making insects, as may be seen in figure 6, numbers 21 and 22, which are of this sort, very closely resemble some of the parasitic species, so that they are very likely to be mistaken for them, yet there are not many of them injurious to farm crops, nor likely to be seen about them. For this reason the insects of the sort under consideration which the farmer is most likely to come across, are his friends. Dragon flies devour many small insects, and in all the great groups into which entomologists divide the immense kingdom of the insects, there are some which, by destroying their allies, become most valuable friends to man.

Somewhat resembling in form and closely related entomologically to the ill-smelling black squash bug, there are several insects which are most beneficial.

Figure 1 shows the form of one of these, and figure 2 an enlarged cut of another, which is popularly known as the Spiny Soldier Bug. It is one of the most useful predatory insects, devouring in the larval state plant lice in great numbers, and when fully grown it attacks a great variety of injurious insects, some of them being among our most destructive spe-

Figure 1.
SOLDIER BUG.
Acanthocephala femorata. Gl.

Figure 2.
SPINY SOLDIER BUG.
Sinea mltispinosa. De Geer.

cies. Even the potato beetle finds an enemy in this insect, as does the canker worm and other pests. Figure 3 represents a fly devouring a larva which he has seized. Flies of this group, known as Asilus flies, are beneficial in destroying pernicious insects, but they, unfortunately, do not confine themselves to those that are harmful, but often destroy many bees. Figure 4, as well as figures 5 and 6, illustrate a large group of small, often brilliant-hued beetles, known as Tiger Beetles, and the name is significant of their habits. They may often be seen running briskly over paths or roads in

Figure 3.
ROBBER FLY.
Erax apicaulis, W.

Figure 4.
TIGER BEETLE.
Tetracha carolina, L.

Figure 5.
TIGER BEETLE.
Tetracha Virginica, Hope.

bright warm days. They devour many insects and are to be regarded as beneficial.

Figure 6 gives illustrations of several species of the Tiger Beetles and the larva, with the head of one of them enlarged.

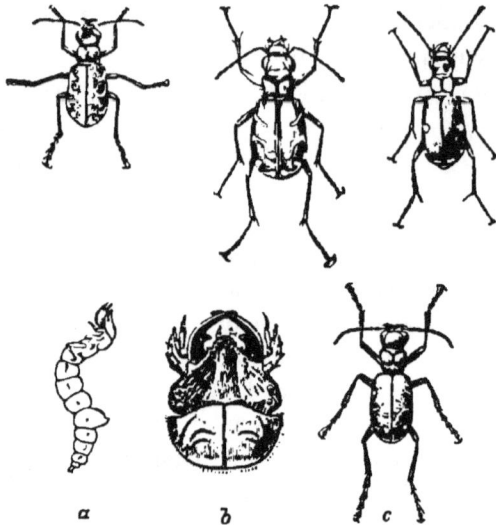

Figure 6.

Cicindelidæ—TIGER BEETLES.

Beginning at the upper right hand corner the species are: *Cicindela hirticollis, Say; Cicindela vulgaris, Say; Cicindela sexguttata, Fabr.* a. larva, b. head enlarged, c. beetle. (After Riley.)

Figure 7 shows the larva and perfect insect of another and larger beetle of similar habits. It is not very uncommon, and is exceedingly useful in destroying some of our most pernicious insects. The beetle is bronze-black, with light reddish bronze spots arranged in rows, as seen in the figure. It is found under stones and in other sheltered places. Among other insects, it destroys the white grub.

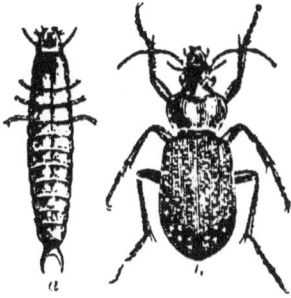

Figure 7.
Callosoma calidum, Fabr.
LARVA and BEETLE.

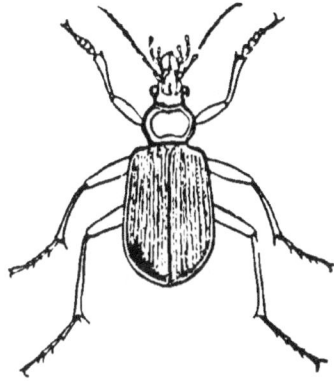

Figure 8.
Callosoma scrutator, Fabr.

Figure 8 illustrates a larger species of this beetle, of similar habits. Dr. Harris commends it for destroying canker worms, even going into the trees after them. If these beetles could be in some way increased so that they should become abundant, it would be a blessing to the farmer.

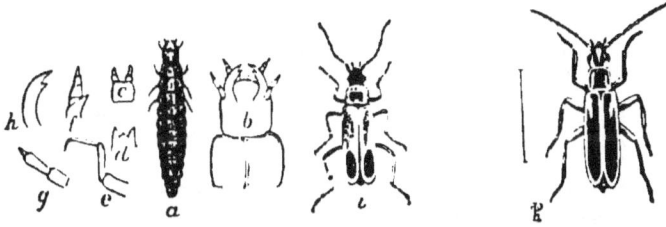

Figure 9.
Chauliognathus Pennsylvanicus.
SOLDIER BEETLE.
a. Larva; *b.* Larva, head enlarged; *i.* Beetle;)
c. d. e. f. g. h. Parts about the head.
13

Figure 10.
Chauliognathus marginatus.

Another group of thin-winged beetles are serviceable in the same manner as those named. These are sometimes called soldier beetles, though the significance of the name is not very apparent. Figures 9 and 10 show the form of these beetles. Figure 10 is not found in this region. The larvæ feed upon the larvæ of many insects; that of figure 9, among others, eats the larva of the plum *curculio*. Few beneficial insects are more useful than the so-called lady-birds or rose beetles. They belong to a large family known as the *Coccinellidæ*. These insects will be more fully considered later in this paper, since they are especially valuable as devourers of plant lice.

They are frequently seen, often several of them together, about our windows in early spring, and have more than once caused needless alarm because they were mistaken for the dreaded buffalo bug. Several species are seen in figures 59, 56, 57. They feed upon plant lice in both the larval and adult state.

Figure 11, upper figure perfect beetle, lower larva, shows a large and not very uncommon beetle which is especially useful as a devourer of cut worms, although it does not limit itself to this diet, but destroys the larvæ of other insects.

Figure 11.

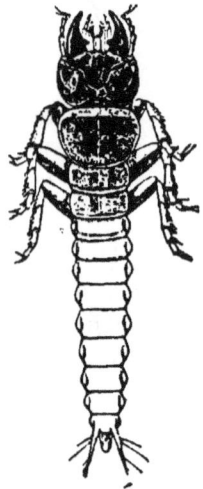

Figure 11.
Harpalus caliginosus, Say.
BEETLE and LARVA.

Wasps and hornets are disliked by fruit growers because of the injury which they often do to fruit, and they also sometimes destroy bees, but they destroy great numbers of injurious insects, and are to be regarded as highly beneficial. Figure 12 represents a common Wasp. There are many species of very minute flies or fly-like insects, which lay their tiny eggs on or in those of injurious insects, and the egg of the parasite hatching before that of the

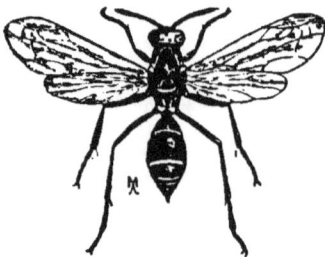

Figure 12.
Polistes bellicosus.

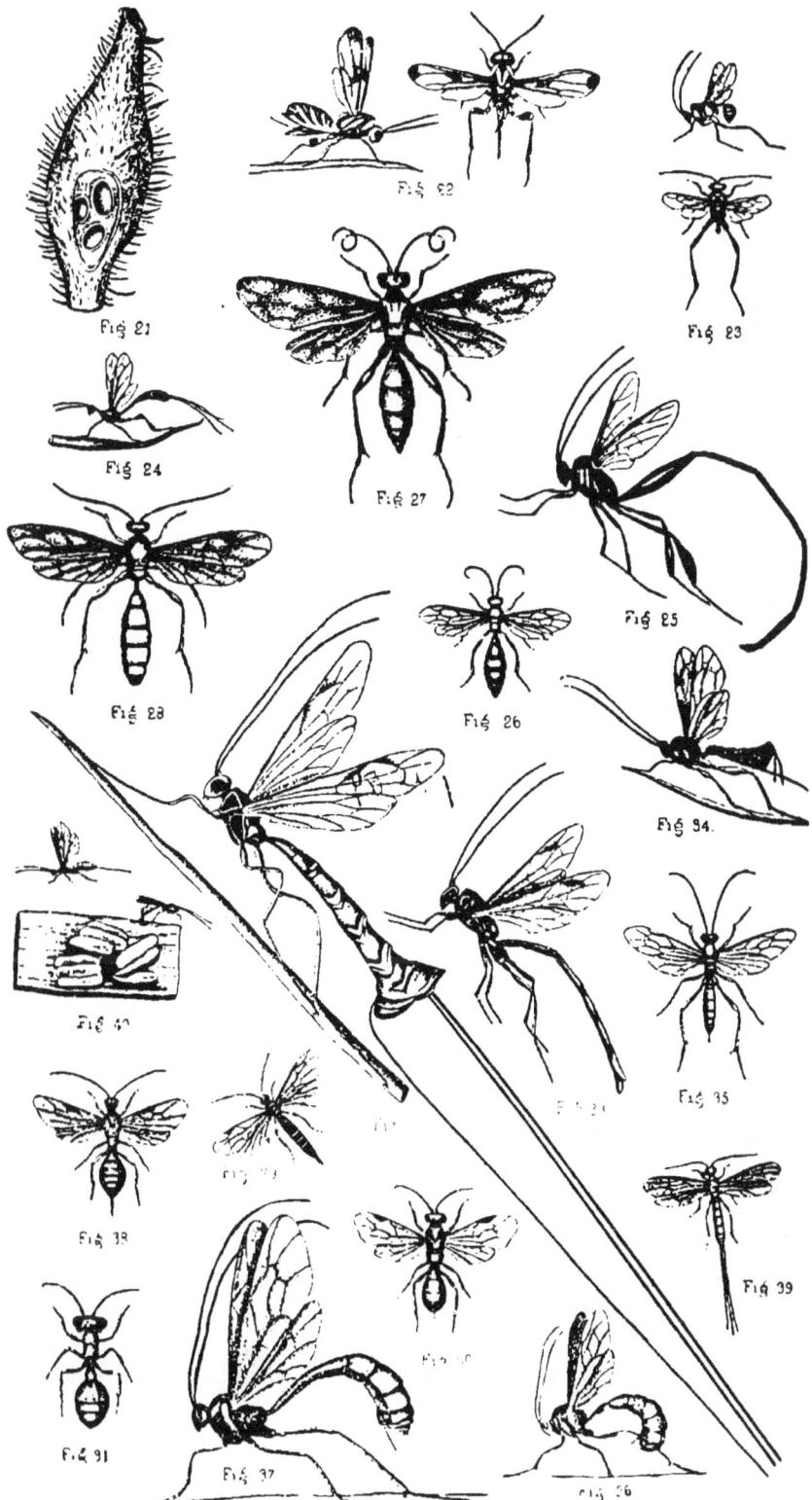

Fig 21

Fig 22

Fig 23

Fig 24

Fig 27

Fig 25

Fig 28

Fig 26

Fig 34.

Fig 40

Fig 35

Fig 38

Fig 39

Fig 91

Fig 37

Figure 13.

host, its larva, thrives upon the surrounding food. They are not true flies, since they have four wings and belong to the same group that contains the bees, ants and wasps.

Most of the figures given above are copied from the papers of Packard, Riley and Comstock.

The full-page plate noted as figure 13, gives numerous examples, taken from a paper by Townsend Glover, of insects, which it is well for the farmer to know.

Figures 21 and 22 are gall-making insects, but figures 23 and 24 are predacious insects, though European ; figure 25, an ichneumon, *Pelecinus polycerator*, is not uncommon in New England, and I have several times found it in Burlington. It is probably a beneficial insect ; figure 26, *Ichneumon suturalis*, and figure 27, *Ichneumon grandis*, 28, *Trogus exesorius*, figure 29, *Cryptus-inquisitor*, all are parasitic. The eggs of the ichneumons are laid on larvæ or pupæ of various injurious insects, which are thus destroyed, sometimes in large numbers. Figure 30 is a European insect, and also figure 31, both parasites. Figures 32 and 33, . *Rhyssa lunator, Fab.*, female and male, are parasitic on the *Tremex columba*, which is described later as infesting the elm. It is wonderful how these insects are able to thrust their long, slender ovipositors into the trees after the larvæ of the Tremex upon which its larva feeds. *Pimpla pedalis*, figure 34, is a parasite on the tent caterpillar. Figure 35 is a European insect ; figure 36, *Ophion macrurum, Linn.*, is a parasite on one of our largest larvæ, that of *Telea polyphemus* ; figure 37, *Ophion bilineatus, Say*, attacks the common yellow bear caterpillar ; figure 38 is a European insect ; figure 39 is a Southern species, which, as does that given in figure 40, lays its eggs in the holes made by borers. It will be readily understood that I do not expect the farmers to be able to identify the insects mentioned as well as other family species by these figures, but it has seemed to me that a series of illustrations, such as these given, might prove very helpful in giving farmers some idea, even though it be only a general one, of the general appearance of some of the insects which devour injurious species. While not strictly correct, yet the statement is sufficiently so to be of value, that most of the common insects which have the form and appearance of those given, are beneficial rather than injurious, and to be treated accordingly. As may have qeen noticed, of the parasitic insects some of them lay their eggs on the eggs of the injurious species, sometimes on the larva, sometimes on the pupa. Figure 14 illustrates the larva of one of the grapevine moths covered with cocoons of a parasite. Figure 15 shows the adult form of the parasite, both natural size and enlarged.

Figure 14. **Figure 15.**

Figure 16, from Packard, shows a fly which is beneficial. It is known as a Tachina fly. These flies are parasitic in caterpillars and other insects. They lay their eggs on the caterpillar, and then hatching, the maggots or larvæ, destroy their nest. I once obtained thirty-nine of these flies from the cocoon of a single Cecropia, and many a collector has been disappointed in trying to obtain a specimen from the cocoon when, instead of the expected moth, he only secured a small swarm of these hairy flies.

Figure 16.
TACHINA FLY AND LARVA.

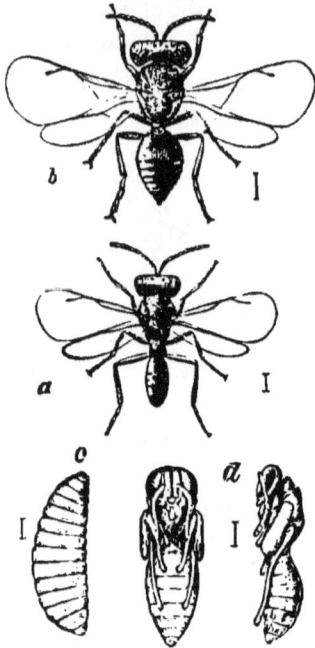

Figure 17 illustrates another parasitic insect, which is a foe of the cabbage butterfly.

Figure 18 gives another parasitic insect belonging to the Ichneumon tribe. It is destructive to the larvæ of some of the larger moths.

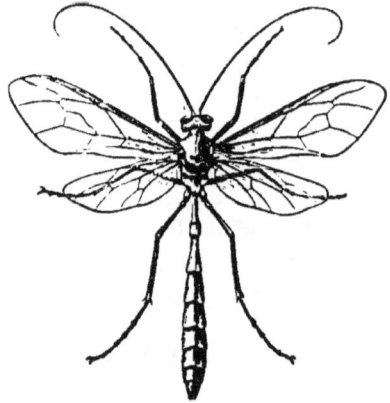

Figure 17.
Pteromalus puparum, L.
a. Male; *b,* Female; *c.* Larva; *d.* Pupa.

Figure 18.
Ophion macrurum.

II. OTHER ENEMIES OF INJURIOUS INSECTS.

Birds and toads are also of the highest value as insect destroyers. As to birds, their value is more and more fully recognized, and consequently those species which are totally insectivorous are more and more welcomed and protected by the farmer. And as for toads, if anyone will take the trouble to watch them for a time, he will be speedily convinced of their value. It is true that these friends of the farmer are not always as discriminating in their choice of food as is desirable, and too often devour beneficial as well as injurious insects, and on this account their aid is not wholly desirable. Still a very large balance remains to their credit, after all deductions have been made, and he who drives away or destroys either our common insect-eating birds or the toads is making a great mistake. Various quadrupeds, as skunks and raccoons, devour many insects, but probably none are so valuable as the common pig when he has free range. It is a most judicious and beneficial plan to turn swine into a field which is infested with insects which live in any of their stages in the ground, or which infest fruit which falls to the ground.

III. INSECTICIDES.

Every farmer or fruit grower has long ago discovered that he cannot leave the care of his insect foes to their own special enemies, but must enter the list himself and seek out the best modes of attack. Especially during the last few years has the study of insecticides attracted attention, and many new substances have been tried, some of them very useful, some of them of little value, as proved by ample experiment. Other subtances are yet on trial, and at present it is wise to speak guardedly concerning them, while many new insecticides will undoubtedly appear from time to time. One cause of failure in the use of insecticides has been the improper application of that which in itself was of value. A knowledge, the more complete the better, of the feeding habits, time of appearance and of the different periods of the life of each insect, is essential if the use of insecticides is to be profitable. In more than one instance considerable sums of money and great labor have been thrown away because of mistake in the identity of the insect to be destroyed, because of which remedies useful against one insect were used against another of so different habits that the remedy was wholly unavailing. A notable case of this sort happened some years ago in Baltimore. From the nature of the case no insecticides can be of universal application; some are indeed much more generally useful than others, but none can be satisfactory in all cases. Moreover a remedy highly useful when applied at a certain time, may be of no value whatever when used at another time. The time and method of application are often of the greatest importance. Hence when our fruit or crops are infested, the first thing to do is to discover what is the nature of the destroyer, its habits, mode of growth, etc., and these being known, we may then go on to devise and apply a remedy with some chance of success. If the farmer himself is in doubt of these matters, he is always at liberty to make his wants known at the State Experiment Station, and, so far as possible, the Entomologist of the Station will give all needed information.

14

ARSENITES.

Of the numerous substances now in use, probably none are more generally useful than the Arsenites, Paris green and London purple. There is still some difference of opinion among entomologists as to the relative merits of these two poisons, and it is quite probable that each has its advantages over the other, one being better in some cases while the other is best in other cases. I have experimented somewhat for several years with both, and find that London purple is on the whole better than Paris green for most purposes, since it is cheaper and quite as effective. These may be used as dry powders diluted with plaster, flour or some similar material, but generally they are more useful when used with water sprinkler or sprayed on the plants or trees. The fact needs to be kept in mind that no solution can be obtained by putting either of the substances into water, only a mixture, and that this may be uniform, not too weak to kill the insects at one time, and so strong as to destroy the foliage at another, it must be frequently stirred, else the poison settles at the bottom of the vessel containing the mixture. As to strength of the mixture, this will vary somewhat with circumstances, but most experimenters agree that a mixture of one pound of London purple and two hundred gallons of water is strong enough for all ordinary purposes and not so strong that it destroys or even greatly injures the foliage. Of course a small quantity could be made in the same proportion, that is, an ounce of purple to twelve gallons of water. Some would use a stronger mixture than this, but probably it would rarely be advisable to use more than one ounce in nine gallons of water. If a second spraying is necessary it should be made with a weaker mixture, that is, one ounce of London purple may be diluted with fifteen or twenty gallons of water. Paris green is heavier than London purple and stronger, containing about ten per cent. more arsenic, but is used in equal bulk. Since the latter is lighter, much more arsenic would be furnished the mixture than the difference indicated would give. Mixtures as strong as one pound of poison to one hundred gallons of water have been used without injury to foliage, but at other times very extensive damage to foliage has followed. According to Prof. A. J. Cook more damage to foliage results from using the poisons late in the season than if used earlier, and "used just before a rain are more harmful than during a drought." Various experiments show that very little of the poison from sprayed trees falls on the grass beneath, and that there is little danger to animals feeding upon this grass. By far the most satisfactory method of using the mixtures is to spray the trees with them by using any good force pump. A small pump can be bought for from two to six dollars. This will spray a few trees thoroughly if not too large, but for an orchard or any extensive operations a barrow furnished with cask and pump, or still larger a cask on a wagon or truck drawn by horses, will be needed. Several large pump manufactories make a great variety of pumps for spraying trees. If fruit trees are to be sprayed this should not be done until after the flowers have fallen, or there will be danger of destroying all the bees in the immediate neighborhood. There is more difficulty in getting large trees and especially fruit trees, such as plum trees, well sprayed so that enough of the poison adheres to the leaves or fruits to be of use. Whenever this difficulty appears, it may be, at least in part, obviated by mixing about a pound of common wheat flour with every twelve or fifteen gallons of the liquid.

KEROSENE EMULSION.

Next to the above in value probably the various kerosene emulsions should be placed. The object of all of these mixtures is to secure a thorough commingling of the oil with the other liquids so that no free oil shall be found. Very few plants can stand a dose of clear kerosene, although some are not injured by a very moderate amount. I have never sprayed a plant with clear oil and do not expect to do so, but I have used oil drawn from the barrel in one or two desperate cases of mealy-bug, using a soft brush and applying the oil in this way. It must always be a case of kill or cure when such a remedy is used, but it was cure and no appreciable injury in the cases above mentioned. I have also used a mixture of kerosene and carbolic acid, adding many times their bulk of water with good results. but I have no doubt that the kerosene emulsions would have been better, although the mixture used involved much less trouble. Various recipes of making the emulsion of kerosene have been tried and all are good. Prof. Riley's recipe is a mixture of oil and milk of any desired proportions, although that recommended is one gallon of milk, sour is better than sweet, though either will do, and two gallons of kerosene. These must be thoroughly mixed. Very much of the success of the mixture depends upon this. It needs to be well shaken or churned until a sort of butter is produced. This process may require but a short time or it may need nearly an hour. The warmer the liquid is, up to the boiling point, the sooner the churning is effected. If a stronger mixture is needed the amount of oil can easily be increased in the next lot. In the form of kerosene butter the mixture may be kept indefinitely. When used water is added, from ten to fifteen times as much as the amount of the emulsion used. This can then be put on the plants with a force pump or sprinkled with a watering pot. An emulsion which is perhaps more easily prepared than any other, and is also quite as efficient, if not more so, is one recommended by Prof. A. J. Cook. This is made by taking one quart of soft soap or four ounces of bar soap, and dissolving it in two quarts of boiling water and when the soap, is dissolved and while the solution is hot, a pint of kerosene is poured in. The whole is then vigorously stirred until, when at rest, no oil rises to the top. When used, enough water is added to make fifteen pints, and then it can be applied to the foliage or stems of plants. This can also be used beneficially to destroy lice on domestic animals, although, by some, tobacco water is preferred. I have used an emulsion made substantially like the above on rose bushes with excellent effect. Bushes treated by being thoroughly syringed with the above, using, however, about twice as much water for the final dilution, have been kept free from slugs, and all the other pests so common on roses, while adjoining bushes unsprinkled were left with only skeletonized leaves. A single application may be sufficient, but perhaps it would be better to repeat it a week or two later. If the bushes are well syringed as soon as insects appear fine fresh foliage may be kept as long as the flowers last.

PYRETHRUM.

The Persian insect powder or Pyrethrum is sometimes very valuable and sometimes quite useless. It is often used dry, dusted over the infested plants, but it is easy to see that when used in this way any insectical is much less likely to remain where it is need than if applied wet, and on all accounts it is better to apply it as a mixture with water. A large teaspoonful stirred in a gallon of water makes a good mixture, although for some purposes it may be made stronger. Some of the experiments tried with this mixture have been very successful while others have not been so at all. There are numerous varieties of the powder and some of the failures have undoubtedly been due to weak powder. It is not at all poisonous, and hence is free from some of the objections which some persons may make to the arsenites; but at best it is not so certain as they are. The pyrethrum has the advantage that it is not necessary that it be eaten; it destroys insects, if it does so at all, by its contact with them. But on this account it is necessary that it be very thoroughly applied, that it may at once reach all the insects. For the green cabbage worm and other soft-bodied insects it is to be recommended.

CARBOLIC ACID.

A dilute solution of carbolic acid (the crude dark sort which can be bought for about seventy-five cents a gallon is good enough) is very good as a remedy for plant lice, mealy bug and such like insects. The strength may vary according to circumstances; but it cannot be used very strong or the foliage will be killed. One part of acid and one hundred of water makes about as strong a mixture as it is safe to use. An emulsion may be made with carbolic acid instead of kerosene, and this is less likely to injure the foliage. It may be made stronger than kerosene emulsion. Prof. Cook's formula, which is certainly good, is as follows: One part of carbolic acid to five or six parts of soap solution. As a remedy for bark lice, borers, and the like, this is very highly recommended. It is to be rubbed over the infested trees, or those liable to be infested, about two weeks after they are in bloom. When it is desirable to use a dry mixture, lime, plaster or any powder may be charged with carbolic acid, one part in fifty well mixed, and then dusted over the plants. Few classes of injurious insects are more difficult to manage than those which remain in the ground the greater part of their life. The white grub is of this sort, and many remedies have been proposed and tried with more or less success, and for the most part the success has been less.

BISULPHIDE OF CARBON.

At present the most promising substance which can be used for the extermination of such insects is bisulphide of carbon. This is a very volatile, ill-smelling liquid, exceedingly inflammable, but not dangerous in any other way. Its value consists in its readily becoming a vapor, which penetrates the soil immediately about the the spot where the liquid is poured, and in the deadly effect which this vapor has upon insects. It is most easily used, all that is necessary being a hole in the infested ground made with stick or bar, into which a little of the liquid is poured and the hole closed by a lump of earth. The liquid, as has just been noticed, speedily becomes a vapor, which extends on all sides for some distance, varying with the nature of the soil, and kills all insects with which it comes in contact. It has been used in Europe in large quantities and with great success for killing root-infesting insects, and of late has been used to a less extent in this country. It may also be used on carpets in closets, or anywhere the housekeeper may discover moths, always providing that no spark of fire shall come in contact with the vapor and that the disagreeable odor can be tolerated. This odor, however, is not without value, for by it the presence of the vapor and the danger from fire can readily be detected. After using it a thorough ventilation removes all danger, and after all it is little, if any, more dangerous than benzine, and more efficacious, though more expensive. In pound bottles it costs about thirty cents, but it is said that in large quantities it can be bought from the makers for about twelve cents. It is not, as hitherto used at any rate, useful against leaf eaters or for spraying, since it is injurious to the foliage and not as certain to kill as the arsenites. This substance is also useful, and carbolic acid and kerosene can be used in the same way, to destroy ants when they become troublesome, by punching a hole in the hill, pouring the liquid into it, and then plugging the opening.

HELLEBORE.

I know of no better agent to use for currant worms, rose slugs, and the like, than the long-tried white hellebore, but as a general insecticide it is not so valuable as some of the others named above. It is very often used as a dry powder, but, like pyrethrum, and indeed all insecticides of which I have any knowledge, it is more easily, and thoroughly, and consequently, more effectually applied as a liquid. One ounce of the hellebore, as purchased, may be mixed with four or five quarts of water, and then be sprayed or sprinkled over the bushes.

COPPERAS.

A solution of copperas in water has been used with good results, in some cases, and as it is so simple and easily procured a remedy, it is worthy of trial. It is of greatest value when used on soft-bodied larvæ, but it may well be tried. The strength need not be great ; two ounces in a quart of water would be sufficient.

Much less than this does not destroy many insects and perhaps it had better be stronger rather than weaker. A long list of substances and mixtures might be given in addition to those mentioned, but none of them are of so great value as those given, and most are of doubtful efficacy. Some other substances have had, and still have, more or less reputation among farmers, which are of no value whatever, so far as I can discover, and of course their use involves loss of time and permits the uninjured insects to go on with their ravages. I refer to such remedies as buckwheat flour, fine, dry wood dust, etc. These have been thoroughly tried to no purpose, except the establishing of the fact that they are useless. I have tried these and other substances for cabbage worms more than on any other plants, but I have never seen a worm destroyed or even driven away by them, and yet they have been very confidently recommended time and again. A remedy for plant lice, mealy-bug, etc., which I have tried thoroughly, is in itself so simple that it seems almost absurd: it is simply keeping watch of plants, and when insects appear directing a small stream of water against them until they fall off. For six years the plants in a conservatory have been kept free from the common pests of such a place in this way. In watering by hose nozzle or force pump, all that is needed is that the opening be partly stopped by the finger, in order that a small, forcible stream may be obtained, and this is directed upon infested plants. The shock of the stream so far disables the insects that when once knocked off they rarely have strength to crawl to the leaves again, but if they do the next watering finishes them. I do not see why small shrubs and plants out of doors could not be treated in the same manner. Of course I only recommend this method as useful in case of non-flying, soft insects, such as those named. It is not supposed that it is of any other than limited application, but it has proved very efficient. No very great force of water is needed, since most insects are easily dislodged by a direct stroke, such as the stream would give. I have attempted to free the same sort of plants from insects by fumigating with tobacco and formerly depended chiefly upon this; but, as has been said, for six years nothing but the cold water treatment has been used, and, with far less trouble, the plants have been kept cleaner than in the old way.

STARVATION.

One method of getting rid of troublesome insects always remains when all else has been tried. In case of fruit and other trees starvation is impracticable; but when field crops are attacked, as a last resort this may be tried. That is, the infested field may be allowed to lie fallow for a year or two, in order that there may be no food for the insects. Sometimes a simple change of crop answers the object. To be successful, all farmers in a given locality should act in concert, else starving out on one farm would be of no avail, if the pest were allowed to increase or even exist on those joining, from which the farm freed by starvation would be speedily stocked.

HYDROCYANIC ACID GAS.

Another insecticide has been used with some degree of success in California, where scale insects have been at times very troublesome. By this method only small trees, or those of no great size, can be treated. It consists in covering each tree with a tent made for the purpose and fumigating it with Hydrocyanic acid gas. As described by D. W. Coquillett in *Insect Life* as follows: "It consists in using one part by weight of dry or undissolved potassium cyanide, with one part of sulphuric acid and two parts of water. The generator is made of lead and is somewhat of the form of a common water pail. After the tent is placed over the tree, the necessary quantity of dry cyanide is placed in the generator, the proper quantity of cold water is added, and the generator placed under the tent near the trunk of the tree; the acid is then added to the materials in the generator, a sac thrown over the top of the latter, after which the operator withdraws and a quantity of earth is thrown upon the lower edge of the tent where it rests upon the ground to prevent the escape of the gas. After the expiration of fifteen minutes the tent is removed and placed over another tree." This treatment does not, if care be taken as to quantity of material, etc., injure the foliage or fruit, but does wholly destroy the insects and at no great cost. I have quoted thus fully from Mr. Coquillett's article because the same process is applicable to other trees in other places, and would prove effectual when everything else fails. It should be always used with care and by one who understands fully what he is dealing with, as the gas is exceedingly poisonous, and must not on any account be breathed.

BORDEAUX MIXTURE.

What is known as the Bordeaux Mixture, though not an insecticide, may properly be mentioned here for in combination with poi-

sons it may be both an insecticide and a fungicide. This prepara-
tion is made as follows : Six pounds sulphate of copper dissolved
in six gallons of hot water, four pounds of unslaked lime are slaked
in six gallons of cold water. After the lime is slaked and cool,
pour the two solutions together and add ten gallons of water. This
proves very useful in preventing rot and other troubles due to
fungi, and by adding to the mixture Paris green or London purple,
it becomes an insecticide as well.

Besides the methods already mentioned numerous more or less
successful experiments are being made in different parts of the
country to ascertain the value of various fungi as insecticides. It
has been known for a long time that many insects were more or
less liable to be attacked by fungi and that they were destroyed by
their growth, but it is only recently that any attempt has been made
to use this fact and to cultivate the fungi in order that they might
be used as insecticides. This has been done thus far only to a
limited extent, but it is not impossible that much more may be
accomplished in this direction. The spores of fungi are very
small, and can be kept for a long time without injury, and when
once their growth is started they develop with great rapidity and
spread with equal facility from one insect to another, destroying
quickly those infested, so that it is possible that a serious outbreak
of injurious insects be very effectually and promptly checked by
means of fungi.

It has seemed to the writer better to devote the remainder of
this paper to a definite group of insects rather than to study various
unrelated groups. Hence we will now turn our attention to

IV. INSECTS INJURIOUS TO THE AMERICAN ELM.

It will soon be discovered that however limited may be the field
of inquiry which the student of injurious insects may adopt, his investigations will necessarily carry him very far beyond the bounds
which his subject may at first seem to draw for him. In studying those species of insects which attack the elm, we shall be
forced, and very readily, to discuss some of the worst pests which
infest fruit trees and other forms of vegetation, so that before we
have completed our task we shall find that the fruit grower and
the farmer may find that which is of value as well as the lover

of shade trees. It should be noticed here that an abstract of the following pages, with some of the illustrations, has already appeared in the recently published Third Report of the State Agricultural Experiment Station, and the illustrations used in the former paper have been loaned by the Station for use in this. No one should regard these papers, although the subject is similar, as duplicates, for not only is the subject treated somewhat differently in each, but this paper is far more complete than the space which could be given in the Station Report permitted. It has been thought that, as full and complete an account as may be of all the insects known to infest the elm would be useful for reference as well as for the present. Of course, in a paper like this very much is necessarily included for the sake of completeness which is not unknown or unfamiliar, to entomologists at any rate, but not a single species of this sort is mentioned, I believe, which has not been mentioned in letters received from different parts of the State and information respecting it desired. So far as the general subject is concerned, it would seem to the writer that no apology or excuse ought to be needed for investigating the enemies of so noble a tree as the elm, and, while some of the species herein enumerated are at present of comparatively little importance, it is yet more important that they should be at least briefly considered than it may at first thought seem to be, because no experience is more common than that a species of insect hitherto inconspicuous, both as to number and damage done by its ravages, suddenly becomes very conspicuous by its rapid and unaccountable increase. No species of insect is to be passed over as of no possible interest, although we may well devote much greater attention to some than to others. From this, it follows that all observations of farmers, fruit growers and all others are of value, and the great advantage which must result from co-operation in the study of insects is manifest. The very great unlikeness which is found to exist between the young and the adult of most insects, renders the study of their life history the more puzzling and gives additional value to every observation. How great the changes through which an insect may pass during its cycle of existence may be will appear by and by, as we study the Aphides in the last part of this paper. But let us return to the elm. As has been said, no excuse ought to be necessary to justify devoting time and space to the insects that are destroying, or may do so, this tree. It is always easy to understand how any damage to farm

crops affects the finances of those cultivating them, and it is quite
natural and proper that those insects which destroy the common
and more important farm crops should have especial attention ; but
it is not right that our attention should be confined to these. Our
shade trees have not only beauty, but active value, so that the most
thoroughly utilitarian and practical man must yet be interested in
their preservation, and there are many who would be very greatly
troubled if the familiar shade trees about their homes, or on their
farms, were to perish. Very fortunately many of our trees, while
not exempt from the plague of insects, are not as liable to destructive attacks as are herbaceous plants. Yet the attacks of insects
are sufficiently numerous and ruinous to call for our diligent study.
There are, as will be seen in what follows, insects often found on
the elms which, if sufficiently numerous, would easily destroy
these trees, root and branch, and if we consider what a melancholy
change would take place in many of our towns and villages if all
the elms were destroyed, we may well feel some anxiety respecting the welfare of these trees, that add so much to the beauty and
attractiveness of our State. I have chosen the elm rather than the
maple, or any other tree, at this time, partly because the elm seems
to me most in danger, and partly because I have been asked to do
so by various persons whose opinion and wishes I very highly
respect. It may seem an over enthusiastic outburst of admiration
to say that the American elm is the finest shade tree known, and
opinions may well differ in regard to this, but for myself I believe
it to be true that, taking gracefulness, ease of cultivation, variety
of form, adaptation to our climate and the physical features of the
country, this tree is superior to all others. There are those that
are more delicate in leaf, more elegant in flower, more wonderful
in this or that respect, but making up a general average, I think
we shall find our elm coming out ahead of all the rest. It is rather
strange that this tree, like most others, has as its worst enemies, for
the most part, very small insects. There are a few large species
that more or less commonly attack some part of the tree, but by
far the larger part of those insects which injure it are small. This
fact makes it all the more difficult to deal with the foes of the elm,
since the large size of the tree and the small size of the insect renders the concealment of the latter easy, and the great extent of surface to be reached by any insecticide makes it difficult to apply it
with sufficient thoroughness to be of use.

Moreover, many of the insects we are to study are insidious in their method of attack, and ere their presence is noticed, they are so far established that their removal is difficult. It is a fact that our elms are in danger; some have already died, some are sickly from the attacks of insects, some less severely affected are yet infested, and on the way to ruin, so that it is none too soon to give attention to them. General remedies have been considered in the first part of this paper, and most of these are applicable to the elm, but it will be better to speak of specific remedies in connection with specific insects. In Europe over a hundred different species of insects have been described by entomologists which infest the European elm. A less number have thus far been found on the American elm, but in the list which is given on a following page, there will be found a sufficiently powerful array of foes to satisfy the most belligerent. In considering the damage which any insect commits, we must have regard to both the time and method of attack. If an insect like the canker worm eats the leaves of a tree, the injury is greater if done in the early part of the season, as it always is, than it would if done later, so, too, since the living, growing parts of a tree are the new roots, twigs and leaves connected by the outer layer of wood and inner layer of bark, an insect which attacks these is more injurious than one which should attack the heart-wood, and so on.

The following list embraces all the different species of insects which, so far as I have knowledge, attack the elm. I am well aware that such a list is not very interesting in itself, but I hope that it may prove of value to some of those who may read this paper. With the scientific name the common name is given so far as possible. Many of the species have no other name than the scientific. In order to aid those unfamiliar with entomology in the use of the list, I have added after the name of each of the large groups under which the different species are enumerated, a few of the more familiar examples in order that some idea of the character of the group may be gained. For convenience I have arranged the names alphabetically.

LIST OF INSECTS INJURIOUS TO THE ELM.

HYMENOPTERA.

BEES, ANTS, WASPS, ETC.

Cimbex americana, Leach. Elm Saw-fly.
Tremex columba, L. Pigeon tremex.

LEPIDOPTERA.

MOTHS AND BUTTERFLIES.

Anisopteryx vernata, Peck. Canker worm.
Anisopteryx pometaria, Harr. Fall Canker worm.
Anthaxia viridicornis, Say.
Amphidasys cognitaria, Guen.
Apatela americana, Harr.
Apatela grisea, Walk.
Apatela morula, G. and R.
Arctia nais, Drury.
Argyresthia austerella, Zell.
Bactra agntaria, Clem.
Ceratomia amyntor, Hub. The Four-horned Ceratomia.
Clisiocampa americana, Harr. Tent caterpillar.
Chœrodes clemitaria, A. and S.
Empretia stimulea, Clem.
Epirritia dilitata, Hub.
Eugonia subsignaria, Hub.
Grapta interrogationis, Hub. Semicolon Butterfly.
Grapta progne, Cram.
Grapta comma, Harr.
Halesidota caryœ, Harr.
Hyphantria cunea, Drury. Fall Web-worm.
Hyperchiria Io, Fabr. Emperor moth.
Hybernia tilaria, Haw.
Lagoa crispata, Pack.
Lithocolletis ulmella, Cham. Elm leaf primer.
Lithocolletis argentinotella, Cham.
Limacodes scapha, Harr.
Melaneura quercivorana, Guen.

Nepohpteryx undulatella, Clem.
Nephopteryx ulmi-arrosella, Clem.
Orgyia leucostigma, S. and A. Tussock moth.
Ocneria dispar, L. The Gipsy moth.
Parasa chloris, H. Sch.
Paraphia unipunctata, Haw.
Platysamia cecropia, L. Cecropia moth.
Phigalia strigataria, Min.
Paonia excæcatus, A. and S.
Scirodonta bilineata, Pack.
Smerinthus geminatus, Say.
Synchroa punctata, New.
Telea polyphemus, Hub.
Tolype velleda, S. and A.
Vanessa antiopa, L. Mourning cloak Butterfly.
Zeuzera pyrina, Fabr.

COLEOPTERA.

BEETLES.

Cotalpa lanigera, L. Goldsmith Beetle.
Chrysomela scalaris, Le C.
Dularius brevilineus, Say.
Galeruca calmariensis, L. (*Gxanthomelæna*, Sch.) Elm leaf
Beetle.
Graptodera chalybea, Ill. Grape vine Flea-beetle.
Hylesinus opaculus, Le C.
Holotrichia crenulata.
Lachnosterna fusca, Frohl. May Beetle. June Bug.
Lachnosterna micans,
Lachnosterna hirticula, Harr.
Magdalis armicollis, Say.
Monocesta coryli, Say.
Neoclytus erythrocephalus, Fabr.
Phloiotrobus liminaris, Harr.
Phyllophaga Georgicana, Gyll.
Phyllophaga pilosicollis, Knoch.
Polyphylla variolosa, Hentz.
Saperda lateralis, Fabr.
Saperda tridentata, Oliv. Elm Borer.
Saperda candida, Fabr.

Sinoxylon basilare, Say.
Trichestis tristis.

ORTHOPTERA.

GRASSHOPPERS, CRICKETS, ETC.

Œcanthus niveus, L. Tree cricket.

HEMIPTERA.

BUGS, PLANT LICE, ETC.

Callipterus ulmifolia, Monell.   Elm leaf Aphis.
Colopha ulmicola, Fitch.   Cockscomb gall Louse.
Gossyparia ulmi, Geoff.   Elm leaf Aphis.
Mytilaspis conchiformis, Bark Louse.
Pemphagus ulmi-fusus, Walsh.
Schizoncura americana, Riley.   Elm leaf gall Louse.
Schizoncura ulmi, L.
Schizoncura rileyi, Thomas.
Tetraneura ulmi, Elm gall Louse.

As has been already stated, some of the insects enumerated do little or no harm, but most of them are possibly injurious and many of them very positively so. The only exceptions are a few which live in dead wood. A few of those given in the list are not found in New England, and several others have not yet appeared in Vermont, at least, so far as I know, although some of them may be found here at any time. The order given in the list will not be followed as the different species are taken up more or less in detail, since it is more convenient to adapt a more scientific arrangement.

SAW-FLY OF ELM AND OTHER TREES.

Cimbex Americana, Leach.

This saw-fly is shown in figure 17, copied from one of Dr. Riley's. The figure shows the different stages of the insect, *i.* being the adult male, *f.* the cocoon, *d.* the larvæ first hatched, *e. e'* the full grown larvæ, *a.* leaves of willow showing location of eggs, *c.* a single egg much enlarged, *j.* shows the curious saw-like appendage with which the female cuts slits in which the eggs are placed, *k.* showing the end of the saw still more highly magnified, *g.* shows the pupa in the cocoon, while *h.* shows the same taken from the cocoon, *b*. shows a twig with slits girdling it, cut by the saw of the perfect insect. The larvæ feed upon the leaves not only of the elm, but also other trees, as willows, birch and basswood. The eggs are laid on the leaves in June, as seen in figure 19 *a*. These eggs soon hatch into larvæ which closely resemble those of many moths, such as the cut worms. They feed for about two

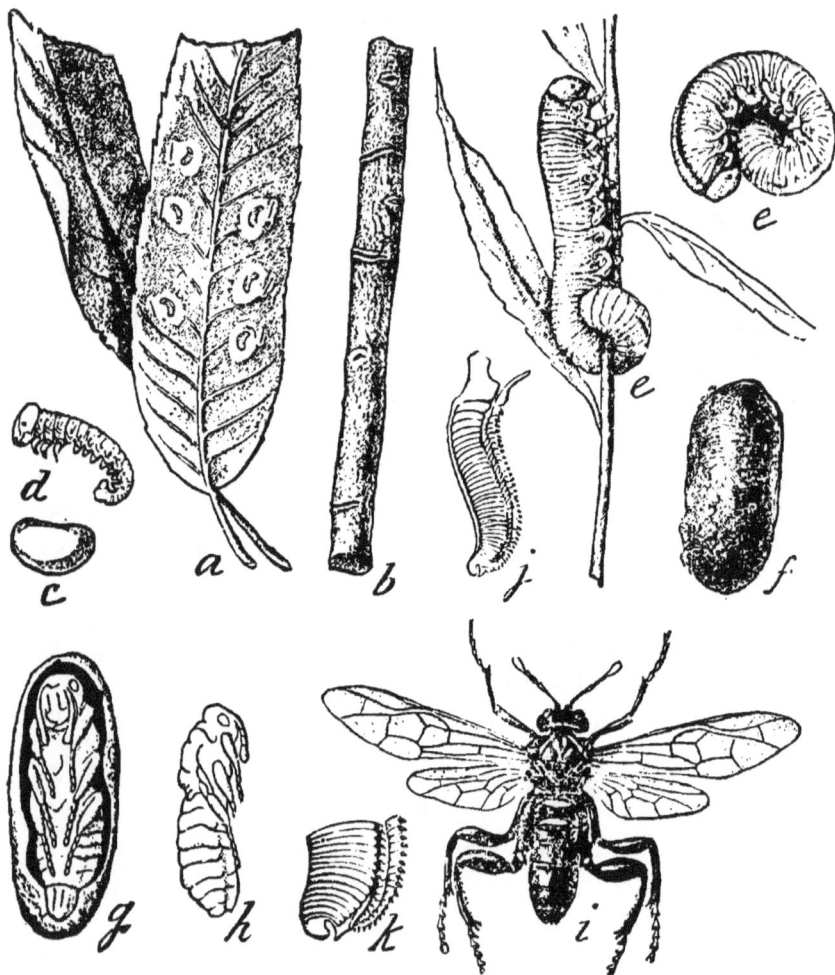

Figure 19.

Cimbex Americana, LEACH.

a. Willow leaves showing
    deposits of eggs.
b. twig girdled.
c. Egg enlarged.
d. Young larvæ.

e. Full grown larvæ.
f. Cocoon.
g. Chrysalis in cocoon.
h. Chrysalis.

i. Adult fly, male.
j. Saw enlarged.
k. End of saw.
(After Riley.)

months, growing rapidly until fully grown, when they are disagreeable looking, greenish yellow worms about two inches long. When at rest they usually curl themselves as at *c*. figure 19. If molested they can throw jets of liquid from openings along the sides, so that all in all they are unpleasant insects. When they have reached their full growth and development they leave the tree on which they have been feeding and descend to the ground, where they conceal themselves beneath any rubbish that is at hand. Having thus hidden themselves, each forms a tough cocoon, figure 19 *f.*, in which it remains during the fall and winter, completing its transformation the following spring. 'The fly which comes from the cocoon is somewhat like a bee as it buzzes about. It is about an inch long, with wings two inches across when extended. The males and females are not precisely alike, for aside from the saw and ovipositor which the female possesses, she is stouter and the body bears several yellow spots on each side, while that of the male is of a nearly uniform bluish color, and he is somewhat larger than his mate.

The size and voracity of this insect make it a formidable enemy whenever it appears in large numbers, which is sometimes the case. Usually, at least this is true in Vermont, it does not seem to increase very rapidly and does no great damage. The currant worm, one of them, *Nematus*, is allied to the above, though a smaller species. It is, however, an insect which has defoliated large numbers of trees elsewhere, and may at any time need attention here. Any of the remedies suggested for leaf-eating insects, such as spraying with kerosene emulsion, Paris green or London purple and water, would be in all probability sufficient safeguards against this insect. Also, taking advantage of the habits of the insect, it would be well to remove leaves and other debris which might be underneath the trees infested after the larvæ had hidden themselves for the winter, or swine turned out so that they had free access to the trees; or even poultry would destroy many.

PIGEON TREMEX, HORNTAIL.

Tremex columba, L.

This is an allied insect, though quite different in appearance from the foregoing. It is larger, and more wasp-like; within the female a spine projecting from the end of the abdomen. The general colors of the female are black and reddish brown, with transverse yellow bands across the long cylindrical abdomen. The body is about an inch and-a-half or two inches long, and the dark wings spread about two inches and-a-half. Concealed as in a sheath in the projection from the end of the abdomen is the borer, a black, needle-like rod. By means of this the insect drills holes into the wood of the elm, and in the holes thus formed the eggs are laid. Besides the elm, other trees are visited.

Figure 20.
Larva of TREMEX COLUMBA.

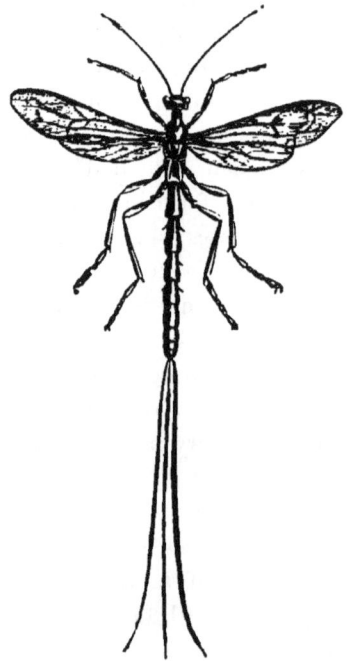

Figure 21.
Rhyssa lunator.

The eggs hatch into a cream colored grub which bores into the wood.

Rhyssa atrata and *Rhyssa lunator*, figure 21, are parasites of the Tremex. By its long ovipositors this insect penetrates the trees infested by Tremex, and places its eggs in or near the burrows, and the larvæ hatching from these destroy the Tremex larvæ. The male Tremex differs from the female, in the absence of the borer, and its shorter body; he also lacks the yellow markings, being reddish brown and black. He is also very much smaller, the expanse of wings being not over two inches, except in rare cases, and usually it is less. The body is usually about an inch long. I do not think that usually very much damage is done by these horn-tails, nor are they very often found in large numbers.

LEPIDOPTERA.

In the next group, the *Lepidoptera*, there are numerous species injurious to our elms, and many of them also injure other trees. It may be well to remind the farmer that the caterpillars, worms, grubs, or whatever they are called, are the first stage after leaving the egg of moths and butterflies, and that by moth is meant not the insect which is such a trouble to housekeepers, though that is a true moth, but the whole great subdivision of *Lepidoptera*. The moths can ordinarily be distinguished from the butterflies by their loving darkness rather than light; whether because their deeds are evil, as they usually are, or from a natural aversion to the sunshine which the butterflies are so fond of, I cannot explain. They also hold the wings when at rest down over the back, sloping from the middle, showing thus the upper side, while butterflies almost always hold the wings vertically, the upper sides inward, thus showing the outside. The antennæ of moths are usually much more complex and feather-like than those of butterflies. In giving the names of the different species discussed, whenever there is a common name by which the species is known, this will be made prominent, but in many cases there is no other than the scientific name, and of course no other can be given. In most cases the larvæ of the moths do more damage than those of the butterflies, but these latter deserve consideration, and all should have some knowledge of their structure and habits. Several species of butterfly attack the elm, though none of them I believe confines itself exclusively to this diet. There are several quite similar species of butterflies quite common in upland fields and the lower portions of mountains, though found everywhere. They all have the upper surface of the wings of a dark, reddish hue, and the underside mottled and dotted with a mixture of white, black, brown, and gray, which reminds one of a cinder. The edges of the wings are variously scalloped and angled. The largest of these is known as *Grapta* (*Polygonia*) *interrogationis*, Fabr., or *Semicolon Butterfly*. The larva of this butterfly is a little more than an inch long, a cylindrical worm striped with red and black. It is covered with black spines. The head is dark red, two lobed. The butterfly appears in May or early June, and a second brood in September or the end of August. The wings on the upper side are shaded with soft, dull red, and mottled with black spots and blotches, which run together near the edge to form a nearly, or quite complete band, outside of which, on the very edge, is a narrow, white band. The under side is dark, ashy gray, very finely mottled. The expanded wings are from two inches and-a-half to three inches across.

Figure 22.
Grapta progne. Cran.

Grapta progne, Cran., figure 22, is somewhat smaller, rarely exceeding two and-a-quarter inches across the spread wings, and usually being less. It is lighter in shade, though of similar color to the preceding. In the former species there is about the middle of each hind wing, on the under side, a silvery mark shaped like a semicolon—whence the specific name. In this species this mark is more like a letter L. The caterpillar is gray. On the front of each segment is a light stripe, and on each side an oblique black spot, and the body is mottled with gray, while the breathing pores along the sides are marked by yellow spots. It is spiny, like the preceding. *Grapta comma*, Harr., also feeds upon the elm. The colors of the wings are much like those of the foregoing, and its size is about the same. The silvery mark on the under side of the lined wings is shaped like a comma, or like the letter c. The caterpillar varies very much in color at different ages, being much darker when young than when mature. The general colors of the mature larva are white, black, and gray, while a line of red spots marks the position of the breathing pores. Besides the elm, the Graptas feed upon the hop, currant, etc. They are more or less attacked by parasites.

THE MOURNING CLOAK BUTTERFLY.

Vanessa (Euvanessa) antiopa, L.

This is an interesting species; it is last seen in the fall and first in spring, since it hibernates in the perfect state, spending the cold months in the nooks of barns and out houses, or other places where it finds shelter. Very few of the Lepidoptera remain through the winter in this state, most of them being in the egg or chrysalis at this time. On account of this hibernating habit the butterflies very naturally are seen late in the fall and even a few warm days in winter may arouse them from their torpor and bring them out of their hiding places for a time. The larva of this butterfly, figure 23, is a rather forbidding worm covered with branched black spines. The body is dark, thickly spotted with white dots and along the back there is a row of eight rather large dark red spots. The head is black. When fully grown the caterpillar is an inch and a half long or more. The caterpillar, though such a formidable looking creature, seems to be perfectly harmless. Probably its array of spines serves it a good purpose when toads or birds are about. They feed chiefly on the willows and elm. The perfect insect, figure 24, is a large hand-

Figure 23.
Larva of Vanessa antiopa, L.

Figure 24
Vanessa antiopa, L.
The right wing shows the under side.

some butterfly in form not unlike the Graptas. The general color of the upper side of the wings is reddish maroon, the border is nankeen, while the spots just inside the border are blue. The under side of the wings, shown on the right of the figure, is mottled with grey, black and white. There are two broods each season. The first comes from eggs laid in early spring, and reaches maturity in July. These lay their eggs, and from them comes a second brood in August or early in September and these last, as has been noticed, remain until the following spring. This species is one of the few found in both this country and Europe. The larvæ are not usually very destructive, but sometimes they are sufficiently numerous to do much mischief. A very interesting fact in connection with this insect is found in the stridulation produced by the wings. Very few of the moths and butterflies are known to make any sound whatever, but it has been noticed in a few species that a more or less distinct noise was produced by them when in company with each other. The sound produced by Vanessa antiopa is a grating or scraping sound, and an examination of the fore wings shows an apparatus for producing the sounds which are made by rubbing the fore wings over the hind.

Of the Hawk moths or Sphingidæ there are three species that have been found feeding on the elm, though none are troublesome as a general thing.

A very fine Hawk moth,

Smerinthus excœcatus, S. and A.

is found in the larva state on many trees, among them the elm. The front wings are richly shaded with fawn color and brown. The hind wings are of a beautiful shade of rose color in the middle, and bear an eye-like spot the center of which is pale blue. The caterpillar is about two inches and a half long, of a green color, obliquely marked on the sides with white or pale yellow. The whole body is covered with light granulations. It is not common enough to be very injurious.

Smerinthus geminatus, Say,

A moth about two inches and a half across the spread wings, is a pretty soft-hued moth of a grayish color on the front wings while the hind wings are very prettily touched with red and bordered with gray. The larvæ is light green, with a curved horn on the caudal portion of the body. The surface of the body is covered with light granulations which roughen it. There are oblique yellow stripes on each side of the body. It feeds on the apple, willow, elm, ash, plum, birch, etc. Another and much larger hawk moth is the

FOUR-HORNED CERATOMIA.

This elegant motli, *Ceratomia amyntor*, *Hub.*, is four or five inches across the wings; it varies considerably in size. The body is large and stout as compared with the wings, as is always the case with the Hawk moths. The wings are rather narrowly triangular. They are of very beautiful shades of brown, gray and white delicately blended. The larva is a stout warty caterpillar usually about three inches long. It is of a bright green color with oblique light lines on the sides, a row of notches along the back. On the third and fourth ring back of the head are two short stout horns, and a longer one on the caudal end. The whole body is covered with granulations. Sometimes instead of being green these worms are brown, and green and brown ones may be seen feeding together. They finally go into the ground to become chrysalids and remain there until the following spring. The large size of the worms renders them often quite conspicuous, and thus they are readily taken and destroyed, but they sometimes do considerable mischief before they are noticed.

Arctia phalerata, Harris.

This is a rather common and very attractive moth seen in summer about our houses. The front wings are velvety black, crossed by a sort of coarse net-work of light buff lines, the hind wings are tinged with red, and bear several black blotches near the hind margin. The body is buff with black stripes. The expanded wings are about an inch and-a-half across. The larvæ is a black caterpillar, hairy, and with warty prominences over the body, and from these arise short white hairs. There is a line of light yellow along the back. It feeds on a variety of crops, and on the leaves of several trees, including the elm.

THE FALL WEB WORM.

This is one of the pests which it would seem easy to discover promptly and remove, but it sometimes gains such a foothold that much injury is done, whole rows of trees being completely defoliated. Although the individual caterpillars are small, their habit of

feeding in company and protecting themselves with a web makes
them often very conspicuous. The adult moth is about an inch
and-a-quarter across the expanded wings. *a*. figure 25, *f*. figure
26. It varies somewhat in color, often being pure white, as in
figure 26, but also often more or less marked with black, some of
the variations being seen in figure 25, which is copied from one of
Riley's.

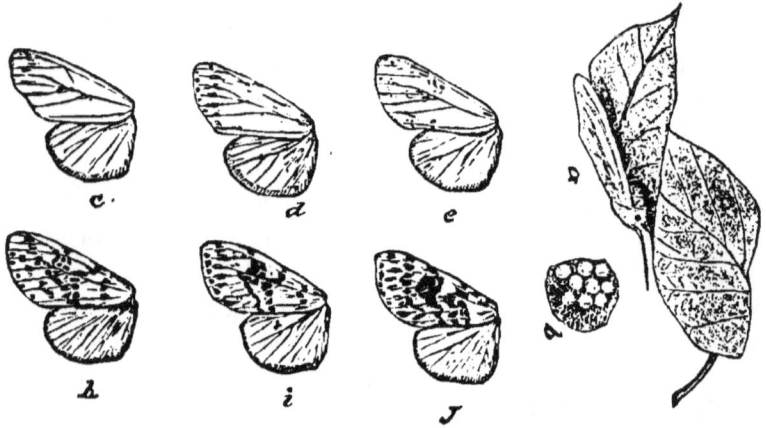

Figure 25.

VARIATION IN MARKING OF WINGS OF FALL WEB WORM.

a. Female resting on a leaf.   *b*. Cluster of eggs.

The females lay their eggs early in the summer on a leaf or
twig. The eggs hatch during July and August ; some of the broods
are early and some are late. Soon after hatching, the little worms
begin to spin silk threads which form the web beneath the shelter
of which they live in companies. These webs may cover a considerable part of a branch. The tent caterpillar uses its web
merely as a shelter at night or during stormy days, and leaves it
during the feeding time ; the web worm differs from it in that it
does not leave the web at all until mature. As the leaves covered
by the web have been devoured the web is extended over new ones.
When the larvæ are full grown they have the general appearance
seen in figure 26 *a*. *b*. *c*. which shows them enlarged somewhat.
As the figure shows, the larva varies in much the same manner as
does the moth in coloration ; some being bright, others dark. They
are about an inch long ; the body is rather thinly covered with
gray hairs, among which are a few black ones. The general color
is yellowish green, dotted or otherwise marked with black. There

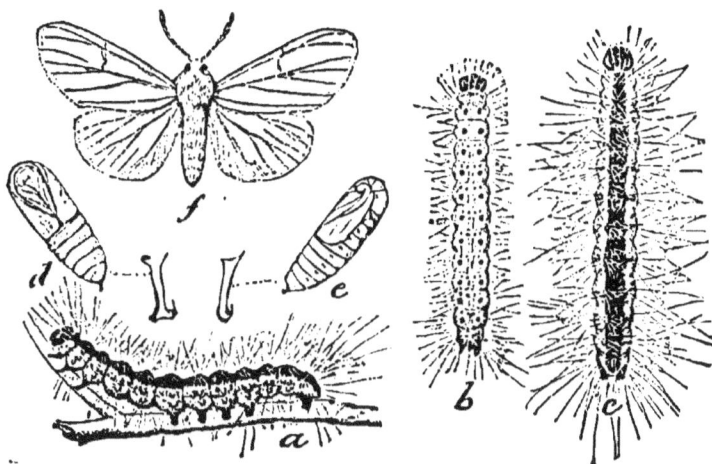

Figure 26.

FALL WEB WORM.

a. b. c. differently marked larvæ.   *d, c.* Chrysalids.   *f.* Moths.

is a yellow stripe along each side. The head is black and the legs and underside of the body are dark. There are twelve prominences or tubercles on each segment, four black ones on the back and four orange ones on each side. They feed on the elm, apple, pear, cherry, walnut, willow, ash, etc. Toward fall they leave the web and enter the ground where they change to chrysalids, figure 26, *d. e.* Of course the most thorough preventive measure in this case is the prompt removal of the web and destruction of its inmates as soon as it is formed, but it may be and often is overlooked for a time; yet even at the risk of mutilating the tree, there is no better way nor any so good as cutting off and burning the twigs included in the web. The usual remedies for tent caterpillar, burning, soaking with kerosene, are just as useful against this insect. but nothing can take the place of watchfulness, and by reason of this the speedy discovery of the web. Spraying with London purple has proved a cheap and effective remedy, and is perhaps the best when the worms have once gotten established upon a considerable number of trees.

Several insects are parasitic on the fall web-worm. Figure 27

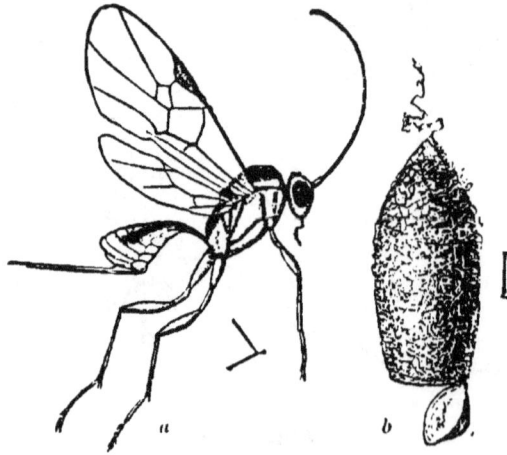

Figure 27.

Meteorus hyphantriæ, RILEY.

shows one of these very much enlarged. Of this Dr. Riley, from
whom the figure is taken, says : " This parasite has performed very
good service during the caterpillar plague, and has done much to
check any further increase of the web-worm." Figure 28 gives

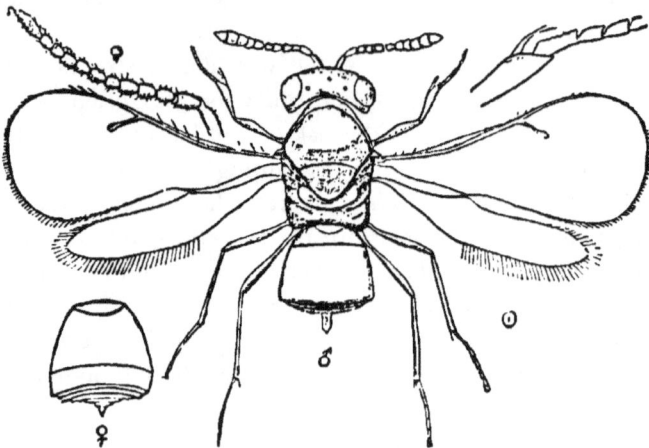

Figure 28.

Telenomus bifidus, RILEY.

another parasite described by Dr. Riley. It is a parasite of the
eggs of the web-worm, and the little insect passes all its stages in

the egg, which is of course destroyed. The minute eggs of the parasite are laid inside those of the moth, and hatch there. Dr. Riley mentions three other species of parasites upon the web-worm.

Halesidota caryæ, Harr.

The caterpillars of this moth are sufficiently abundant to entitle them to a popular name, but I do not know that they have any. Their general appearance is not very unlike that of the next species, though they are less gaily colored. They are thickly covered with white hairs in tufts, while along the back there are eight black tufts and smaller tufts of longer black hairs at the ends of the body. The head and underside of the body are black. These caterpillars feed upon walnut, sumach, ash, elm and perhaps other trees. They are sometimes very common and sometimes rare. They are about an inch and a half long when grown. In the last of summer they leave the trees, go to the ground to hide under stones or in crevices. Here they make a hairy cocoon in which they spend the winter. The next June the moths come out. They are rather prettily marked. They are about two inches across the spread wings. The color is light yellow and the front wings are dotted and marked with brown, and there are three irregular rows of almost or quite transparent white spots across them. The hind wings are very delicate and thin, and bear no markings.

THE TUSSOCK MOTH.

This insect, *Orgyia leucostigma*, S. and A., has done much

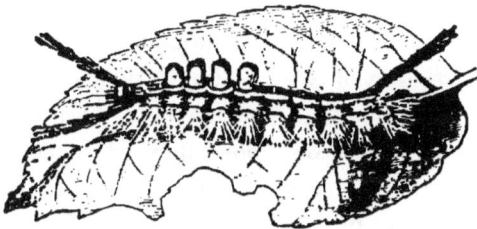

Figure 29.

Larva of Orgyia leucostigma, S. and A.

damage to the elms in some places. The caterpillar, figure 29, is one of the prettiest we see commonly. Like the larva of the

15

moth just described, it is covered, though rather less completely, with tufts of hair of different color ; some tufts being cream color, others black, others yellow. The head is bright red. The general color of the body is yellow. The thick, short tufts of hair seen on the back are yellow or cream color, the long tufts are black and the side tufts are yellow. They usually feed in company and infest various trees ; sometimes they are very common on the apple, sometimes on the elm, sometimes on other trees or shrubs. It seems to vary its habits more than most insects do, for although usually a leaf eater it sometimes eats into young apples, and Dr. Lintner gives an account of a very singular variation in habit which occurred in Albany some few years ago. It also varies greatly in abundance, one year being very common and injurious, and then for a year or several years being quite uncommon. In the instance just named, the Orgyias, which had previously attacked the leaves, left these and began to girdle the young growing twigs too ; so completely did they accomplish this that great numbers perished and were broken off by the wind, so that the walks were thickly strewed with them. Figure 30, copied from Dr. Lintner's Report, shows how the twigs were eaten. It is remarkable that only elms were girdled in this way. Neighboring trees subject to the ordinary attacks of the insects, such as maples and horse chestnuts, were untouched. Dr. Lintner thinks that the Orgyias were unusually numerous because of the English sparrow. This bird does not at all molest these insects, but does drive off other birds that would devour it. In regard to this Dr. Lintner says, in his second Report as Entomologist of New York : " On the sidewalk in front of two buildings, two large spreading elms standing between two maples showed every leaf eaten from them, disclosing the nesting boxes among their branches and their trunks and limbs dotted thickly, or clustered with the easily recognized egg-bearing cocoons of Orgyia. Hundreds of immature caterpillars were creeping over the trees, fences and walks adjoining," and this in spite of the fact that the nesting boxes named were occupied by English sparrows. After feeding for about a month the larvæ spin cocoons, which they attach to some part of the trees or in places not far away. In from ten days to two weeks the moths emerge. The males and females are very unlike, as figures 31

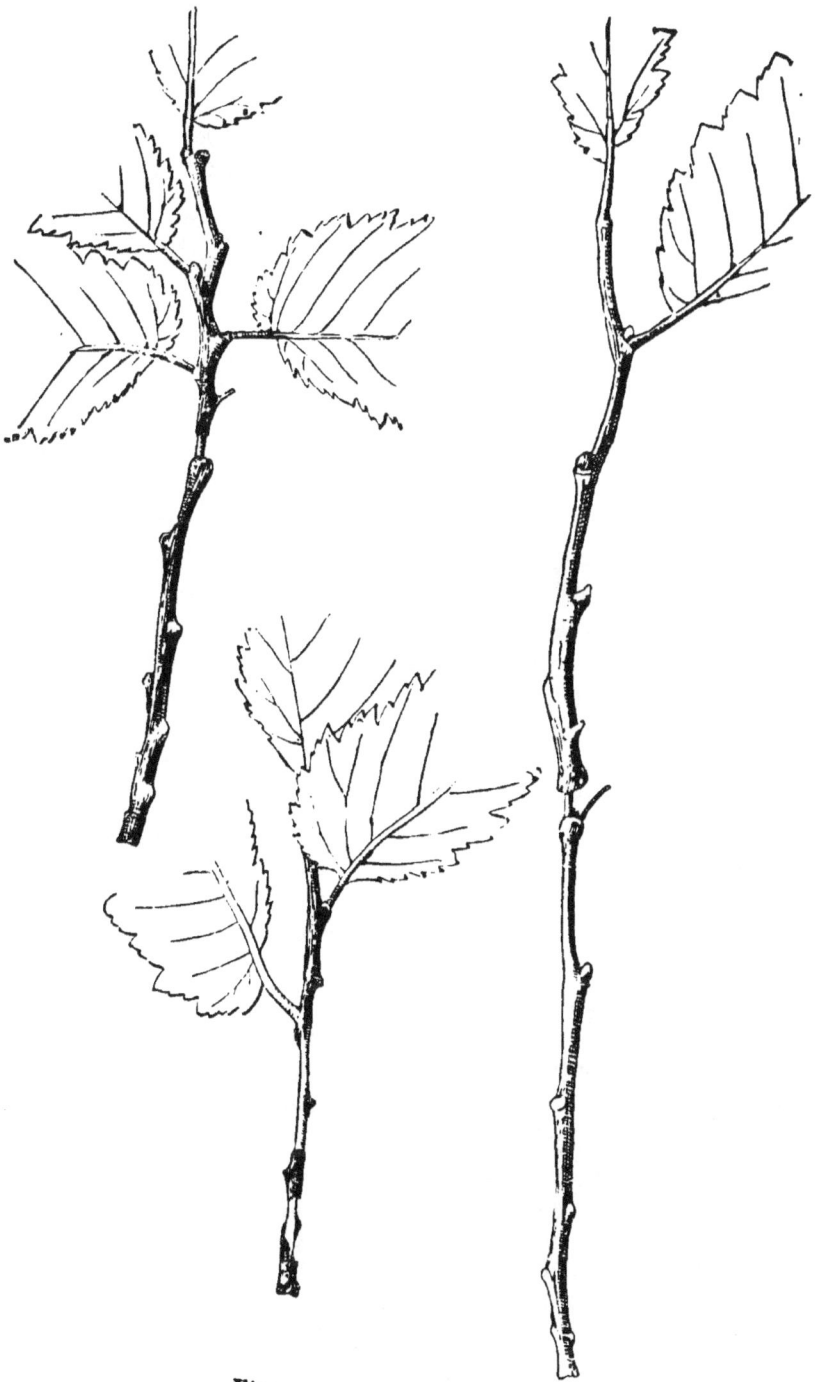

Figure 30.
Elm twigs girdled by Orgyia.

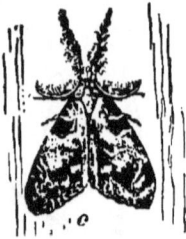

Figure 31.
Male Orgyia.

Figure 32.
Female Orgyia.

and 32 show. The males, figure 31, are like most moths well furnished with wings and they have a peculiar habit of laying back the antennæ, as seen in the figure, and thrusting the fuzzy front legs straight forward. The upper surface of the front wings is in general brownish or grayish with darker markings, while the hind wings, as is very often the case, are of a uniform gray-brown. The females are apparently wingless, of a light gray color. They are of the oval form, shown in figure 32. There are very rudimentary wings which close examination alone reveals. The eggs are laid on the cocoon and covered by a frothy substance, which when dry is white. The worms appear at various times during the season. It appears from Dr. Lintner's observations that the growth of those larvæ which are to become female moths is continued longer than those which become males, nearly a week. The female larvæ molt four times, while the male only three. Since the eggs glued to the cocoons are quite easily recognized, hand picking may be readily resorted to when necessary. The usual remedies against leaf-eating insects can be used with advantage in this case.

The precise scientific name by which an insect is known to entomologists, may not be of very great importance to farmers, and for this reason newer names than some of those given in these pages have not been adopted, since the better known term, so long as it is not really obsolete, is better for our purpose. It should, however, be remarked in passing, that it is not certain that the elm-girdling orgyia are the species *leucostigma* we are considering, but a closely allied species known as *O definata*, Pack. The two species have been very much confused evidently, and may need some overhauling before all is made certain respecting them.

Lagoa crispata, Pack.

The larvæ of this moth appears to feed chiefly on the leaves of

the oak, but it also eats those of the elm. It is a small worm,
when mature about an inch long. The head is white, as is the
body, over which are scattered tufts of slate colored hairs, and there
are also tufts of yellowish hairs. These worms have stinging
power, so that if the caterpillar is handled a sharp pricking sensa-
tion is felt, and the skin soon becomes red. Of this Dr. Lintner,
who has published in the 24th Report N. Y. Museum, a detailed
account of the transformations of this species, remarks, "A critical
examination of the larva, by a partial removal of its hairs, revealed
the existence of clusters of short, slender, acute, white bristles directed upward from the several tubercles of the lateral and subdor-
sal rows, the presence of which had previously been unnoticed,
under their covering of long hairs surrounding and effectually con-
cealing them. Upon touching the bristles with the hand, they
were found to be the source of the sting experienced." The larvæ
make their cocoons toward the end of summer. The moth is of a
dully yellowish red, touched on the sides with drab or slate color.

The stinging property mentioned above is possessed by only a
few caterpillars, and is doubtless to these a very serviceable means
of defence.

Parasa chloris, H. Sch., is reported by Mr. H. Edwards as
feeding upon the elm.

Empretia stimulea, Clem.

A very richly colored moth is sometimes found in the larval
state feeding upon the elm. This has no common name, its scientific name being *Empretia stimulea, Clem.* It is a rare species,
and on that account not especially destructive to the plants upon
which it feeds. The moth is a very handsome, dark reddish brown.
The wings expand about an inch and a half. The larva is a
curiously formed insect, about an inch and a half long, very thick
in proportion to the length, and cut off abruptly at each end. The
general color is a very pretty, bright shade of green, with a large
oval spot of red-brown bordered with a band of white on the back
At each end of the body is a larger and a smaller pair of horns,
which are covered with spines. These have something of the
same poisonous quality as have the spines of *Lagoa* mentioned
above. These larvæ feed on quite a variety of plants, the elm
among others.

Limacodes scapha, Harris.

The caterpillar of this moth would hardly be recognized as such by one not an entomologist, for they have little of the appearance of ordinary species. They are rather repulsive, slug-like, and in the present species boat-like larvæ. The general color is green, with spots of brown. The under side is light colored. The moth is a very neatly marked tan colored or reddish brown insect, with a darker brown semi-heart-shaped patch on the upper side of each front wing, bordered by a light line. The larva feeds upon the leaves of fruit trees and several others, including the elm.

Figure 33.
Limacodes scapha, Pack.

MOTH.   LARVA.

Figure 33 shows this moth, and the curiously shaped larva, natural size.

Scirodonta bilineata, Pack, is reported as feeding upon the elm and buttonwood.   Harris, Ent. Correspondence, p. 301.

The larvæ of several of our large moths feed upon the elm at times, but they cannot ordinarily be considered as very troublesome, for they are rarely so numerous as to do much mischief, and their great size, though this causes them to eat voraciously, yet makes them so conspicuous that they are liable to be seized by birds before they attain maturity.   One of the largest is the

POLYPHEMUS MOTH.

Telea polyphemus, Cram.   The larva of this great moth is not very uncommon, and it has been tried as a silk producer, but without success.   It makes silk enough, but it cannot be wound off and utilized.   A brief account of this insect must suffice.   When mature it is three or four inches long, of a rather delicate green color, with oblique white stripes on the sides.   The moth is five or more inches across the wings, which are in general of a russet brown shading to yellowish.   Near the center of each wing is an oval transparent spot very conspicuous on the hind wings, where it is surrounded by a black and yellow line.   Near the eye spot on the hind wings is a large, beautiful blue spot.   About the lower edge of the hind wings is a light band, above it a dark, and a similar though narrower band crosses the front wings parallel to the outer edge. The whole moth presents a downy appearance, and the colors are beautifully soft. Besides the elm, the larva feeds on maple, hickory, pear, plum, etc., etc.

Our largest and in some respects most elegant moth is the Cecropia.

Platysamia Cecropia, L.

This magnificent insect in soft blending of colors is hardly equalled even among its kindred. The moth is five or six or sometimes seven inches across the wings which are of various shades of grey, brown, red and white. A white band extends across both wings parallel with the the outer border, and in place of the eye-spots of polyphemus there is a little inside of the middle of each wing a large whitish, bean-shaped spot, shaded on the outer margin with dull red, and bordered with black. The margins of the wings are lighter than nearer the body. In Harris' Insects Injurious to Vegetation, may be found a superb engraving of this superb moth. The caterpillar is three or four inches long, of a very pretty shade of green. Each segment of the body bears tubercles from which arise short black hairs or bristles. On top of the second ring these tubercles are bright red, as are those on the third ring; on the seven following rings the tubercles are yellow, and on the eleventh, one large tubercle. Along each side are two rows of light blue tubercles. This is not uncommon in this region, and in the fall its large brown cocoons may be seen hanging to a great variety of trees and bushes. I have taken them often from currant bushes, though they are usually found on larger shrubs or trees.

A much smaller moth, though large as compared with most moths, is what is known as

Hyperchiria io, Fabr.

The larva of this moth is quite a general feeder, eating the leaves of numerous forest trees, as well as a few herbaceous plants. It has stinging spines, which make themselves felt when the insect is roughly handled. Figure 34 gives the general appearance of this caterpillar. As the figure shows, the short stiff spines are in clusters, the clusters being in rows, one about each segment. The

Figure 34.
Larva of Hyperchiria io. Fabr.

color is bright green or sometimes bluish above. The breathing pores along the sides are marked by yellow spots bordered by brown, and below these on each side is a light line. The head is shining green. The spines are some of them yellowish green tipped with black, others are light at the point, some have sharp points, others blunt. Most of them are branched. When fully

Figure 35.
Hyperchiria io, Fabr. male.

grown it is about two inches long. They feed sometimes singly, sometimes in rows. Dr. Lintner mentions finding sixteen " arranged side by side in perfect parallelism," on a leaf of the poplar. They reach their full development about the last of August, and make cocoons in which they remain to appear as moths the next season. The moths, figure 35, male, differ considerably in color according to sex. The males are smaller and of a deep yellow on the front wings, with markings of reddish purple. The hind wings have a border of the same peculiar purple near the body and a narrow band near the outer edge, while there is an eye spot about the middle of each hind wing of dark blue with a white center and a broad black border. The wings are about two inches and a half across. The female is about an inch more across the wings and is of a much darker color, the yellow predominating very little and the general color being a sort of reddish or purplish brown marked with gray. The hind wings are more like those of the male in color but the eye spots are much larger. There is a much larger moth, called by Harris the Imperial moth, which has wings that spread four or five inches, of a yellow quite similar to the male, *io*, and banded with the same reddish purple. Its scientific name is *Eacles imperialis*, Drury. I do not know that its larvæ feed upon elm leaves, and it has, so far as I know, never been reported as doing so, but I suspect it does, for I have taken the moths several times from elm trees, and in parks where no other trees were near at hand and where no oak, walnut, or any of the trees given as those upon which the larva feeds were within a considerable distance. The common tent caterpillar, *Clisiocampa Americana*, is not reported as feeding upon the elm, and I know of but one instance of its doing so. Last year a medium sized elm near my house was infested by this insect, and one quite large nest was allowed to remain till late fall. I have also seen the larvæ upon one or two young elms this summer.

A moth which is not at all common, known as *Tolype velleda*, Stoll., feeds upon the elm as well as upon the apple and oak. The moth is one of the lappet moths, a wooly or fuzzy insect, with white and gray wings, about an inch and a half across in the males, and an inch more in the females. The caterpillar has a bluish-gray body, according to Dr. Lintner, with longitudinal lines. There are scattered parts from which proceed hairs, which are black on the back and gray on the sides. When fully grown it is more than two inches long.

In *Garden and Forest*, January 15, 1890, Professor Smith publishes an account of a new insect which attacks the elms. It has thus far, I believe, been found no nearer than Newark, N. J. It is an imported species, and although it has been in the country for several years, has only within a short time attracted attention. This is a pretty moth known to entomologists as *Zeuzera pyrina*, Fabr. Like the preceding, the female is much larger than the male. Her wings spread two and a half inches. According to Professor Smith, the eggs are laid in the forks of the small branches, and "the young larva burrows downward toward the larger branches, tunneling generally through the center, and usually killing the branch." The larva grows until it is over an inch long. It is a white worm, dotted with black, and from each dot comes a black hair. This insect appears to have gained a foothold about Newark, and to be spreading. The only remedy seems to be to cut off and destroy the infested twigs. Several species of the germs *Apatela* feed upon the leaves of the elm.

Apatela americana, Harr.

This is a gray moth, two or two and a half inches across the spread wings. The general color of the front wings is a soft, rather light gray, marked, as the figure shows, with lines and blotches of darker color. The light marginal line is also seen. There are also spots of lighter color scattered over the surface. The broad wings are of nearly uniform color, but darker than the front ones, especially in the female. The caterpillar is hairy, rather large, being, when full grown, two inches and a half long. Its general color is dark greenish, varying to blackish, the head being brown. Besides numerous yellowish, bristle-like hairs, there are longer tufts of black hair, one near the hind end of the body on the eleventh ring, and two on the fourth, and two more on the sixth ring. The feet are black. It appears late in summer, and does not reach its full size until fall. It then goes into a chink, crevice, or any suitable shelter, and changes to a chrysalis, in which it remains until the following summer. These caterpillars feed upon the leaves of the maple, poplar, chestnut, basswood and elm, and perhaps other trees.

Another species, *Apatela morula*, G. and R., was some years ago described by Dr. Harris, under the specific name *ulmi*, from *Ulmus*, the generic name of the elm, on which Dr. Harris thought it fed. The larva is described by Dr. Lintner in the twenty-sixth Report of the New York State Museum. It quite closely resembles the bark of trees, on which it rests. It is smaller and lighter in color than the preceding. It appears to be a rare species, and Lintner thinks that it feeds only at night. *Apatela grisea*, Walk., is reported as feeding upon the leaves of the elm.

Eugonia (*Ennomos*,) *subsignaria*, Pack.

A very delicate white moth, figure 37, which is about an inch and a half across the expanded wings ; is sometimes quite injurious to the elm.

Figure 37.
Eugonia subsignaria, Pack.

The caterpillar is a slender measuring worm, rather more than an inch long, of a dark brown color, with a reddish head. They by no means confine themselves to the elm, but are quite general feeders. There is no common name ; the scientific name is given above. Another measuring worm is the larva of a very pretty little moth.

Metanema quercivoraria, Guer.

As the specific name, from *quercus*, oak, indicates, this caterpillar is especially fond of oak leaves, but it is also found on the elm. It is a light green worm with reddish markings.

Paraphia unipunctaria, Haw.

Another measuring worm is mentioned by Dr. Packard, as follows : " Eating the leaves in June ; a gray span worm, an inch and a half long, sprinkled with black dots and short lines, its head and neck a little thicker than the body, each ring with a small, squarish white spot above on its hind edge, and with two blackish parallel lines on each side of this spot." The moth is a light brown or fawn colored insect, paler on the under side. It is about an inch and a half across the spread wings.

An uncommon moth, *Chœrodes clemitaria*, A. and S., which feeds in the caterpillar state upon the various plants—also eats the elm. It is a Southern species, and not likely to occur here. The moth is buff or tan color; on the wings dotted with black. The larva is a light reddish, measuring worm, the body being dotted with brown.

An insect which in many respects is like the canker worm, though usually less common is

Hybernia tilaria, Harris.

It is, however, a larger and more showy insect, as figure 38 shows, at least the male is, for the female, like that of the canker

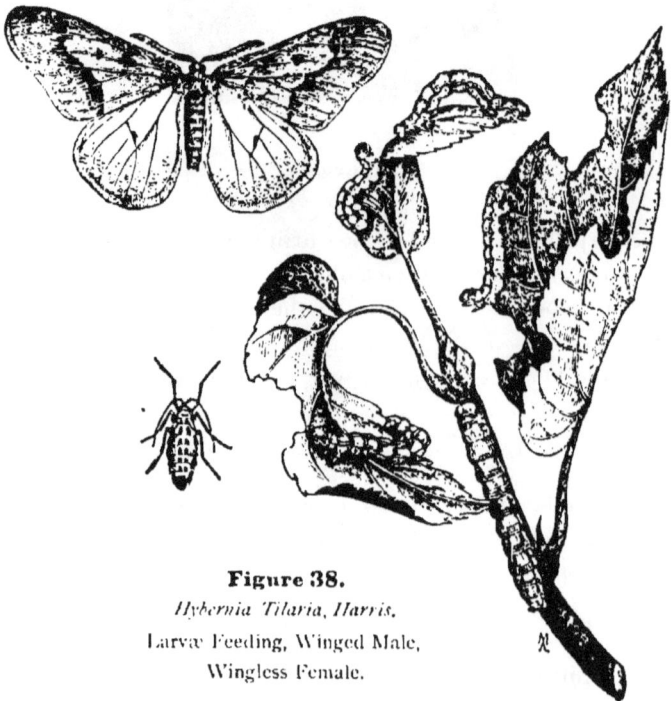

Figure 38.
Hybernia Tilaria, Harris.
Larvæ Feeding, Winged Male,
Wingless Female.

worm, is wingless, and while her mate flies about with entire free-dom, she is compelled to creep in a rather abject manner.

The male (upper figure) is of a yellowish color, shaded with spots of brown, and, as may be noticed in the figure, there is a very brown line across the front wings. The hind wings are brown, and not spotted or marked. The female is of the somewhat oval form shown in figure 38. She is of a light yellow or even white color, with a double row of black spots along the back. The larva is a measuring worm about an inch or an inch and a quarter long. Its body is yellow, with ten longitudinal black stripes, and there are sometimes dark spots. The head is reddish brown, and the under side of the body is yellow. It feeds on the leaves of the basswood, elm, and a great variety of other trees. After feeding in May and June the larvæ descend from the trees by means of a silk thread by which they let themselves down to the ground. They enter the earth and make a cell a little below the surface, and remain as pupæ until the last of October or later, when the moths appear. The female seeks to creep up the trunk of some adjacent tree that she may place her eggs near the ends of the branches, where they may remain to hatch the following spring. The habits of this moth are so much like those of the canker worm, that the remedies recommended for that insect may be applied to this.

This group of geometers or measuring worms to which the in-sects we are now studying belong, is a very large and very interest-ing one.

The caterpillars belonging to it vary a good deal in form, size, etc., and yet they may, for the most part, be readily recognized by their peculiar way of measuring their length along as they creep. Instead of moving with the gliding or undulating motion of most cat-erpillars, these cling by the front or true feet, and loop up the back until the hind or false feet are brought up to the front ones, then the front part of the body is thrust forward until the back is again straight, and thus by alternating looping and extending, they move along. As has been noticed, the group or family of measuring worms is a very large one, embracing hundreds of species. One of these measuring worms is called

Amphidasys cognitaria, Guen.

I do not think it has any common name. The moth is white, "very thickly sprinkled with ashy black," Packard. The larva is a dark brown, or greenish worm, rather thicker than most of the group. The body is dotted with light tubercles, and there are

Figure 39.

Amphidasys cognitaria, Guen.

numerous short, black hairs. Figure 39 shows the moth of this species. The larva feeds upon the leaves of currant and other

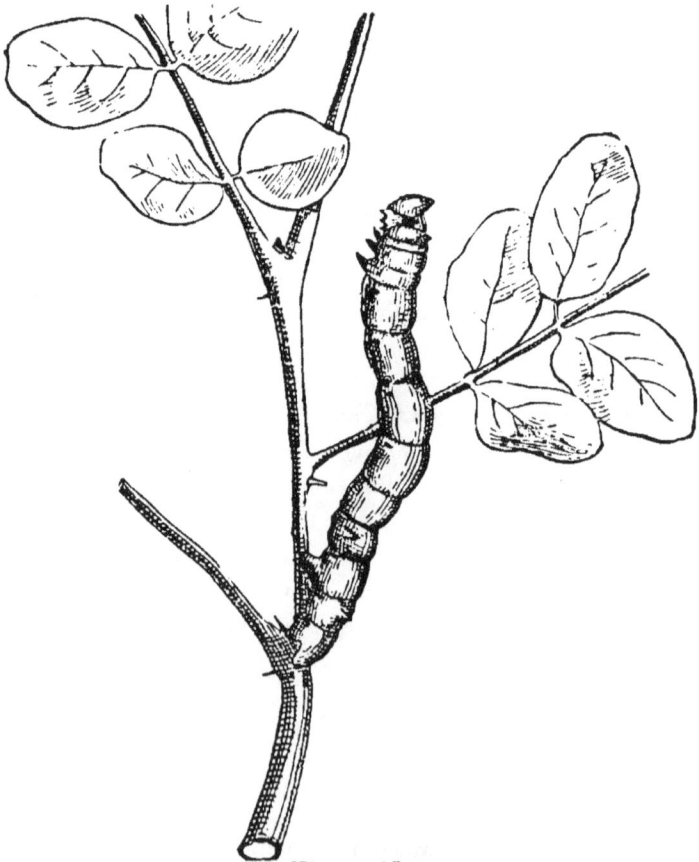

Figure 40.

Larva of Amphidasys cognitaria.

shrubs, as well as those of the elm. After feeding for some weeks, the larvæ descend to the ground, in which it changes to a pupa, coming out as a moth the following spring. The moth is about two inches across the spread wings. Figure 40 shows the larva feeding, copied from Dr. Lintner.

CANKER WORMS.

These familiar, and often most troublesome pests, often commit great ravages upon the foliage of the elm. We have two quite similar species of canker worms, which may be designated as the spring canker worm, and the fall canker worm. Both are very mischievous insects, but the spring species has been most so, and in the many places no enemy of the elm has committed greater apparent ravages, although these are of such a nature that if they are not too long continued, the trees usually recover much of their original vigor in a few seasons.

THE SPRING CANKER WORM.

Aristopteryx vernata, Peck, sometimes occurs in such numbers that the trees are entirely stripped of their leaves. I have seen rows of great elms as bare in June as they ordinarily are in December. Fruit growers, also, are usually made aware of the presence of this pest before they have had many years of experience.

Figure 41.
Anisopteryx vernata.
a. Male.   *b.* Female.

The only really redeeming quality which the canker worm possesses is the clumsy, wingless, and therefore immobile character of the female. A few species of moths are found in which, although the males have fully developed wings, and are very active, the females are destitute of wings and can only creep slowly from place

to place. Had the females of these moths the same well developed locomotive powers as have their mates they would be even
greater and more terrible scourges than now. The sluggishness
of the females limits the range and checks the increase of these
pests so that bad as they often are we may be thankful that they
are no worse. The general form of the male and female canker
worm is seen in figure 41, *a*. the male, *b*. the female. This insect
has been often described and need not occupy great space here. It
will suffice to call attention to the delicate, exquisitely fringed wings
of the male, which are of a satiny luster, and light brown color.
The front pair as seen in the figure, are darker than the hinder
pair and are crossed by more or less distinct wavy dark lines, and
there is a band of very light, almost or sometimes quite confluent,
spots near the outer margin of each front wing. The wingless
female is more or less wholly covered with scales and hairs of a
light brown shade, and as may be seen in the figure there is a
wide black or dark line extending along the back, *d*. figure
42, shows the upper surface of one of the rings magnified ; *c*. is
the ovipositor enlarged ; *c*. a part of one of the antennæ.

Figure 42.
Anisopteryx vernata.
Egg and larva.

In the spring before the leafbuds unfold the moths emerge from
the chrysalids in the ground. The females instinctively creep toward the nearest trees and slowly ascend their trunks, the males
fluttering about them and pairing takes place during the ascent.
If her progress is uninterrupted the female reaches the leafbuds
and places her eggs, figure 42 *b*., greatly enlarged, on or near
them. The eggs are usually laid in clusters. Sometimes the eggs
are concealed in crevices of the bark. The eggs hatch as the leaves
unfold and the young worms at once begin to devour them, growing large as the leaves grow less. About the last of June the

worms reach their full growth and are slender measuring worms, about an inch and a half long, figure 42 *a*. When first hatched the head is glossy black and the body dark green or olive; but during successive molts the color grows lighter, the head is not uniformly colored, but spotted. Along the body there are eight more or less distinct lines or bands of a lighter color than the rest of the body, as shown in the magnified section of the back, figure 42 *d.*, and of the side, figure 42 *c.*, but the markings are far from constant, different specimens showing great variation. When mature the worms creep, or very often lower themselves from the trees by a filament of silk. I have seen trees from which during the last week in June the worms were hanging by hundreds and dropping to the ground. Having reached the ground the worms burrow a few inches, then they turn and work themselves about until a sort of cell is formed and in this they change to pupæ, and remain through the winter, for the most part emerging, as we have seen, the next spring. It is obvious to any one that any method which can be devised to prevent the female from ascending the trees and placing her eggs near the leafbuds will effectually check the increase of the insect, since the young must perish unless they speedily find proper food at hand when they hatch, and most of the remedies which have been proposed have sought to accomplish this. Various appliances have been tried, but none seem more satisfactory than simple bands covered with some sticky substance that will remain sticky in all weather. Probably that recommended by Dr. Packard, cheap printer's ink, is as good as anything and better than most; coarse molasses or tar and oil are good. Mr. Saunders says that if the worms are found on a tree they may be removed by jarring them off in the same manner as that used to get rid of curculias. Spraying with poisonous liquids would destroy many, and when once established on a tree, this method would probably be better than any other. Various parasites attack the canker worm. One, figure 43 which shows it greatly magnified, has often proved most helpful in this respect, so than I have known seasons in which the trees were very seriously affected, to be followed by one in which there was scarcely one to be seen, and the chief agent of destruction was this little parasite, which is one of the tiniest of insets. It lays its eggs, one in each of the eggs of the canker

Figure 43.
Egg Parasite of Canker Worm.
16

worm, and the larva hatching devours the egg substance about it. Many of these worms are devoured by the beetle shown in figure 7. Toads devour many of them; cedar birds and other birds also destroy many. It should be noticed that if bands about the tree are used they should be in place early in the spring, and as some escape from the ground and lay their eggs in the fall it is well to apply them also in October, when they may remain all winter.

The fall canker worm is very similar in appearance to the species just described, and was long confounded with it. Dr. Riley, from whose paper the accompanying cuts are taken, first showed the differences between the species with such fullness and accuracy that they could, in most cases, be easily separated; and the distinction is an important one, because the remedies applied and the mode of application differ somewhat in the two cases, as there is difference in their habits.

THE FALL CANKER WORM.

Anisopteryx pometaria, Harr., is found, not only in the fall, but throughout the season in one form or another. The moths of Anisopteryx vernata, though generally appearing in the spring, may appear in the fall, or even in winter during several warm

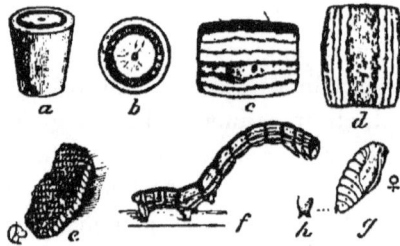

Figure 44.
Anisopteryx pometaria, Harr.
Eggs, Larva and Chrysalis.

days, but the *Anisopteryx pometaria* only appear in the fall and, after laying their eggs, perish. This latter species is not so generally injurious as the spring canker worm, for they are not so general feeders, confining themselves mostly to the elm, although they feed also upon apple and pear trees. Although the eggs are laid at different times, they hatch at about the same time in the spring. The egg of the fall canker worm differs very greatly in form and appearance from the other species, as a comparison of figure 44 *a*. and *b*., with figure 42 *b*. shows at once. They are laid in larger and more conspicuous clusters, and are not often concealed, as are those of the common canker worm. The larvæ are less readily distinguished, although they can generally be separated. As seen in figure 44, *c. d. f.*, as compared with figure 42, *c. d. a.*, the markings along the back are less numerous. The head is darker in the fall canker worm, and the larvæ when first hatched are lighter green. They feed alike on the leaves of the food plants as they unfold in the spring. The first described species in all cases, I believe, does greater damage than the other, as it is most abundant.

As shown in figure 45, *b*., the female is wingless, the body is smoother and rather more slender than in *a*., vernata, and has not the dark median band. The general color is a dark gray. Figure 45 shows the magnified surface of the back, and this may be compared with figure 45, *d*., to show the difference in the two species.

Figure 45.
Anisopteryx pometaria, Harr.
FALL CANKER WORM.
Adult forms, *a*. male, *b*. female.

The male, figure 45, *a*., of the fall canker worm, is darker than that of the other species, though closely resembling it. Other differences, as well as many resemblances, will be noticed on comparing the figures of one species with those of the other. So far as preventive measures go, I wish to notice that although so similar, the two species require dissimilar treatment. Says Dr. Riley in this connection: "In brief, all the more important measures to be pursued in our warfare against the spring canker worm, such as hindrances to the ascending of the moths in the spring, the removal of loose bark, and keeping the trunk and limbs as smooth and clean as possible, the employment of hogs and fall plowing are in the main useless as directed against the fall canker worm, which must be fought principally by traps or barriers applied to the trees in the fall to prevent the climbing of the moths, which mostly issue at that season." (Riley, 8th Report, p 18.) It is all the more important that when either of the canker worms appear, it shall be at once met by insecticides and preventive measures because of the slow progress which the female is able to make from one tree to another, the rapid increase of the pest being thus an impossibility, and if it can be destroyed in the limited region in which it has established itself, the whole adjoining region may be free from it for a very long time.

The larva of a very elegant moth known as

Epirrita dilutata, Pack.

devours the leaves of many of our forest trees, and is at times a foe to the elm. The larva fig is a measuring worm of a greenish color, lighter underneath, and there is a light line along each side. These are often purplish on the wings. It is a subarctic species, and not likely to be common here. The moth is of a general grayish color, dark and light shades are very finely intermingled. The wings have a spread of rather more than an inch and a half. Chambers in a paper "On food plants of Tineina, (Hayden Bulletin IV. 117,) says of *Argyresthia austerella,* Zeller, I am convinced, feeds in some way on it, (American elm,) and in the latter part of May and in June the imago may be found about the trees. *Bactra argutaria, Clem.,* is reported as feeding on the elm.

There is a group of very small delicate moths, which in the larvæ state burrow in the leaves of various plants. They are very handsome little insects. The moths are long and narrow, and the hind wings are finely fringed. The larvæ feed especially upon the middle part of the leaves, and sometimes, indeed, often the presence of the miner is betrayed by a folding or crumpling of the leaf, although if the mine is mainly on the under side of the leaf, it does not fold, but is simply discolored. The little larva spends its life in the burrow, and transforms into a pupa there. Two species infest elm leaves, known respectively as *Lithocolletis ulmella, Cham., Lithocolletis argentinotella, Clem.*

Ocneria dispar, L.

a moth well known in Europe, has appeared in Massachusetts, and is the subject of a special Bulletin by Prof. Fernald. It has a wide range in the Old World. It appears to be quite a varied feeder, sometimes leaving the trees and attacking the vegetables in gardens. Thus far it is confined to a limited area in Massachusetts, and as vigorous efforts are being made to destroy it, we may hope that it will spread no further. The moth, commonly known as the Gipsy moth, is of considerable size, being sometimes two and a half inches across the wings, although it varies considerably in size, and is often much less than this, while the males are nearly an inch less than the female. The wings on the female are light yellow with brown scalloped lines crossing the front wings, and a row of black spots along the outer edge. The hind wings are of uniform color, except at the edge, where there is an irregular band or row of blotches. The eggs are placed on the branches of the trees in late summer, and hatch the next spring into caterpillars, which feed upon the leaves of trees, and finally become nearly two inches long, of a dark color, marked with light yellow, and there is a line of yellow spots along the back ; along the sides are long tufted hairs.

COLEOPTERA, OR BEETLES.

This gigantic group very naturally includes quite a number of species that infest the elm. Some of these are among the worst enemies of this tree ; others, though more or less injurious to the elm, are much more so to other trees or plants. As was shown in the first part of this article, many beetles are to be regarded as friends, since they destroy many injurious insects. The common and sometimes very troublesome

GRAPEVINE FLEA BEETLE,

occasionally attacks the elm. This little beetle *Graptodera chalybea*, *Ill.*, is injurious not only in the larval, but also in the perfect, state. As they come from their hiding places in early spring, the beetles attack the unfolding leaves. They remain but a short time, but the eggs which they deposit on the leaves soon hatch, and the larvæ continue the work of devastation with increased vigor.

Figure 46 shows the various stages of this beetle, and its bright, metallic blue wings and body are sufficiently familiar, especially to vine growers to render a detailed description unnecessary. Air slaked lime dusted over the infested leaves, kerosene emulsion, arsenical mixtures properly, are generally effective remedies.

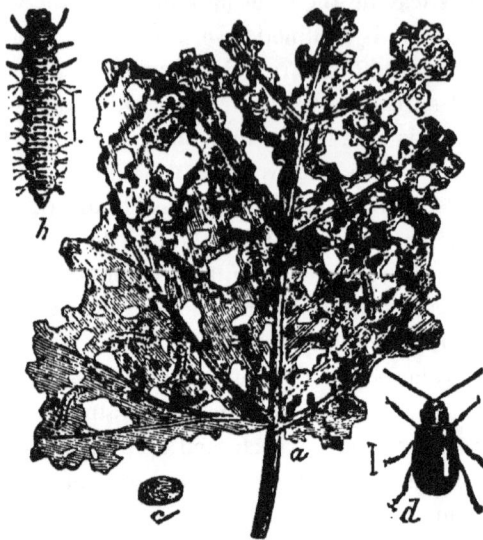

Figure 46.

Graptodera chalybea, Ill.

a. Larva feeding; *b.* Larva enlarged; *c.* Pupa cell; *d.* Beetle enlarged.

Cotalpa lanigera. L.

This is a large golden-bronze beetle, not usually found except in the southern part of the State. It does not do very much damage to the elm. The larva of this beetle is much like the common white grub and, like it, feeds in the ground. It attacks the elm and other trees only in the perfect state.

MAY BEETLE, JUNE BUG.

This beetle, the larva of which is the too well known white grub, is much less harmful to the elm than to those plants which it attacks in the larval state by devouring their roots, yet when the beetles are as abundant as they are sometimes, they destroy no small number of young leaves of our common shade and fruit trees as they fly about them at night in May or June. This beetle *Lachnosterna fusca, Frohl.* was described, and remedies suggested in the Second Report of the Vermont Experiment Station, but as no figures are given in that paper, they are given here from a cut obtained for the Third Report. The figures are natural size and give a very good idea of the insect. It is important that as many of the beetles be destroyed as possible, for this will tend to

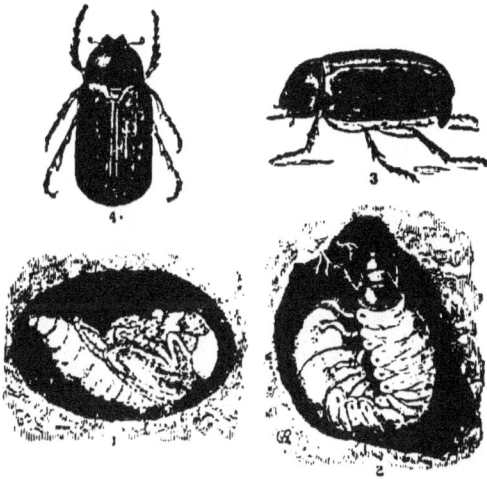

Figure 47.
MAY BEETLE.

1 Chrysalis in the ground.   2. larva feeding.   3 and 4. beetles.

lessen the number of white grubs, which is the form shown as 2 figure 46.   Where the white grub is abundant in the ground it is often very difficult to reach it by any remedies, so much so that some of our best entomologists have recommended starvation by leaving the land untilled for one or two seasons.   In the last Report of the Experiment Station a new remedy has been recommended which, I have strong hope, will prove efficacious and not very costly, although further experiments are needed to determine just the best method of application.   This remedy is bisulphide of carbon, a liquid in some respects like benzine or gasoline, in that it readily vaporizes, and the vapor is destructive to insects.   It is very evil smelling and very inflammable, so that those using it need to keep continually in mind the fact that no spark of fire can be allowed to come in contact with the vapor, if it does, explosion and injury is likely to follow.   In using it, holes should be made in the ground with a small bar of wood or iron, a teaspoonful or two of the fluid poured in, the opening stopped with a lump of clay or anything of the sort.   The volatile nature of the bisulphide causes it to change rapidly from fluid to vapor and penetrate the surrounding earth and destroy such insects, white grubs, wire worms etc., as may be within reach.   Just how far apart and how deep the holes should be, must depend upon the nature of the soil and other local conditions, and must be determined by experiments in different localities, but in general it may be said that they need not be very deep, for the vapor is heavier than air, and will go down as well as up, and as a rule the white grub feeds only a little below the surface. As to distance apart, the holes should be not more than three or four feet from each other but the wisest course would be to try a small bit of ground first and see what the effect is. In Europe the bisulphide is used by the ton for the destruction of phylloxera on grape roots, and with very good success. Should the demand for bisulphide increase as I think it certainly will, it will be possible to buy it much cheaper than now. At present in one pound cans or bottles it costs about twenty-five cents, but it can be bought in larger quantities from the manufacturers for almost half this or even less.

Mr. W. B. Alwood makes a report in Insect Life, vol. 1, page 48, on the use of kerosene emulsion for the white grub. The emulsion used was made by taking ten gallons of oil, and five pounds of soap, which when diluted with water made three hundred gallons. " This was applied liberally to the soil, which was for some days kept freely soaked with the mixture." No serious injury to the grass resulted, and the grubs were effectually driven away, or destroyed. This is a very simple and easily applied remedy, and is worthy of consideration.

In a paper on the " Food and Habits of Beetles," published by Mr. Townend Glover, in Reports of Department of Agriculture, 1868, *Lachnosterna micans*, *Lachnosterna hirticoila*, *Holotrichia crenulata*, *Trichestes tristis*, are mentioned as feeding upon the foliage of the elm. Another somewhat similar beetle is one known as *Polyphylla variolosa*, Hentz., which is found near the coast, but is not common in New England inland. It is a light brown beetle, about an inch long, with very singular fanshaped antennæ. On the back are light blotches. It is found in July. Dr. Harris mentions as feeding upon the elm, *Phyllophaga georgicana*, Gyll., and *Phyllophaga piliosicollis*, Knoch. Among the leaf-eating beetles none are more destructive than those called

THE ELM LEAF BEETLE.

This is a small beetle which has attracted considerable attention for some years. It is said to have been brought from Europe, on imported elms in 1837, but for a long time it did not receive much notice. Of late, however, the ravages committed by the beetle have been so great that its habits have been carefully studied. In general form and size this beetle, *Galeruca xanthomelæna*, Sch., resembles the common striped cucumber beetle. The wings are yellow, varying in shade from light to dark. On the head and thorax are black spots, and on the wing-covers black lines, a narrow one along the inner edge, and a wider along the outer. The under side is black, and the legs are yellow.

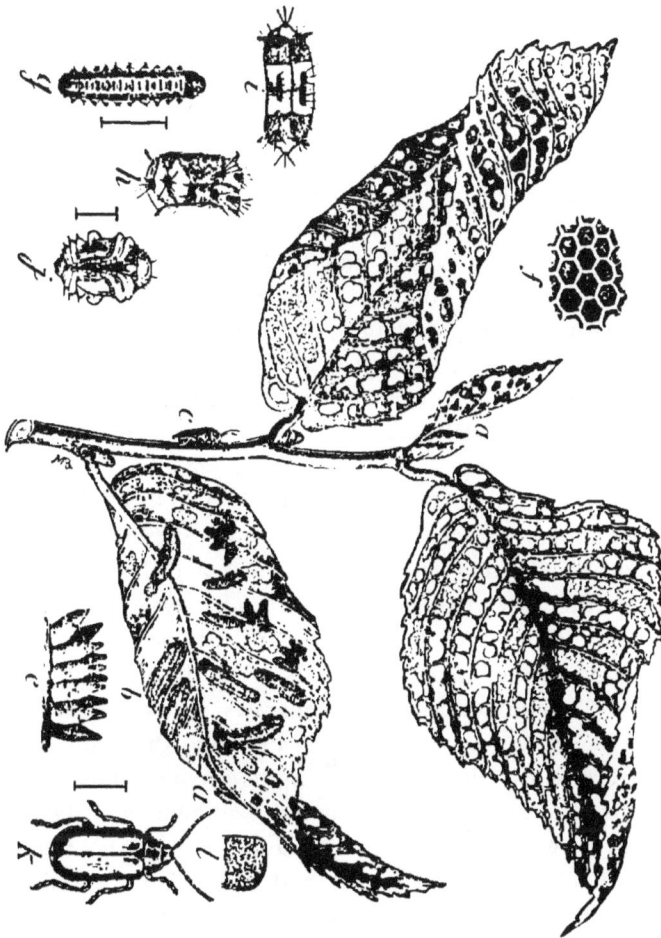

Figure 48.

ELM LEAF BEETLE.

Galeruca xanthomelæna, Sch.

a. cluster of eggs; *e.* eggs magnified; *f.* surface of eggs highly magnified; *b.* larva, natural size; *g.* larva enlarged; *h.* joint of larva, side; *i.* joint of larva, back; *c.* beetle, natural size; *k.* beetle enlarged; *j.* pupa.

The eggs, figure 48, *a*, are laid on the under side of the leaf, where, if all is favorable, they hatch in about a week into hairy, yellowish black larvæ, figure 48, *b. g.* This is in May, or early in June. The larvæ grow brighter in color, as they grow older. When fully grown they are about half an inch long, and of the form shown in the figure. As seen in the enlarged sections, shown at *h.* and *i.*, the hairs are in tufts. Beside the yellow band along the back, there is a similar one along each side. The head is mostly black. After hatching, the larvæ at once begin to feed upon the leaves, eating out the pulp, leaving the woody framework as seen in figure 48. They continue feeding only about two weeks, but as the eggs do not all hatch at once, the trees may be infested by them at any time from May to August, although the length of the time during which they feed would vary with season and temperature. When fully grown the larvæ go to the ground and change to pupæ. According to Riley, the pupa stage lasts from six to ten days. This insect bids fair to become one of the most dangerous foes of the elm. Concerning its ravages in New York, Dr. Lintner remarks, Fourth Report of State Entomologist, 263 : " During the last few years it has extended its ravages to Long Island and Westchester County, N. Y., where by its complete defoliation of large and beautiful elms, and by the myriads of the disgusting larvæ swarming on the trunks of the trees, it became a common object of observation and execration." It has been gradually working its way northward, and though not yet a pest in Vermont, it does occur here, and may easily become a serious pest. The beetle has been carefully studied at the Department of Agriculture, and figure 48 is taken from one drawn by Dr. Riley. The insect has sundry enemies among beetles, which keep it in check to a certain extent. Plowing the ground under infested trees when the larvæ are in the ground, thus exposing them to birds or swine, which should be done after the first of August, is to be recommended. Spraying the infested trees, has on the whole, proved most useful, and has, I believe, uniformly produced good results. The mixture which, after very numerous experiments by different persons in different parts of the country, seems best, is the following : Three-eighths pound London purple ; three quarts flour ; forty gallons of water.

The object of the flour is to render the liquid more adhesive so that it does not so readily wash off from the leaves and twigs. In the United States Agricultural report for 1883, pp. 167-169 there is given a detailed account of the methods employed at Washington to destroy this beetle and the results obtained. These were, in briefest statement: "From midsummer until autumn the unpoisoned half (of a grove of elms) remained denuded of foliage while the poisoned half retained its verdure." As has been stated the beetle was imported with the European elm, but it has gradually extended its operations to the American elm. One of the most helpful accounts of operations against this beetle which I have come across was published by Prof. J. B. Smith, in *Garden and Forest*, June 19, 1889, and as the suggestions therein given apply almost or quite as well to any other leaf-eating insect and to any other tree, I quote several paragraphs. "The trees on Rutgers College Campus were effectually protected by spraying with a mixture of one pound of London purple in a hundred gallons of water. In this proportion the larvæ are destroyed and the foliage not injured." Six pounds of common wheat flour was added to the above mixture, and in order that it might adhere better to the downy underside of the leaves, one gallon of common kerosene emulsion was added. This mixture proved very satisfactory for all purposes. Concerning methods of applying the above Prof. Smith, says: "The results of an application of this mixture are quickly noticeable. The young larvæ succumb almost immediately, and many of the eggs are destroyed where they are fairly hit by the mixture. Another point to be observed is the time of application. This should be just after the eggs are hatched, and before the larvæ become half grown. One application will probably be sufficient, though the tree may be somewhat eaten by the imago. If the trees to be protected are small and few in number it will pay to spray twice, once when the beetles are beginning to deposit and again when the eggs are hatching. If sprayed too early a few beetles will be destroyed which will be replaced by later arrivals and little good is done. If sprayed too late the advanced larvæ will be ready for pupation and will not be affected. Many a man has tried poisons to destroy insect pests and declared them to be failures; they were not so because of any fault of the poison, but simply because they were put on at the wrong time and in the wrong way. It is advisable to avoid wetting the trees more than necessary. The finer the spray, and consequent coating of poisonous mixture, the better the results." The elms referred to by

Prof. Smith were large trees and the whole top could be reached only by means of special appliances. A force pump and wheeled tank were used; to the pump was attached a hose fifty feet long at the end of which was, of course, a nozzle. This was attached to a bamboo pole ten feet long and by this means trees could be sprayed from the ground to a height of twenty or thirty feet, and with a ladder all parts of the largest trees were reached. Of this Prof. Smith says : " Few shade trees are larger than those sprayed by me and no more apparatus would ever be required. For the largest trees, over fifty feet in height, I use about twenty gallons of water containing about one fifth of a pound of London purple and one pint of kerosene emulsion, at a cost of seven cents and the labor of applying the mixture. The result has been to destroy all the beetles and larvæ and most of the eggs, and has preserved fine looking green trees instead of skeletons with fragmentary patches of withered leaves." The importance of using the right means in dealing with insects is shown by an incident in the history of this beetle which happened some years ago in Baltimore. The elms of that city were being greatly damaged by some leaf-eating insect which was hastily supposed to be the canker worm, and the city authorities incurred a considerable and useless expense in protecting the trees against an insect which was not there, instead of first making sure of the insect, and then using proper means for its destruction. The insect in this case being the Elm Leaf Beetle, and not the canker worm, the remedy did little good.

A very pretty little beetle is

Chrysomela scalaris, L. Conte.

Like the foregoing, this is a leaf eater. The upper wings are silvery white, spotted with green, while the thin under wings are red. The under side of the body is dark green, as in the thorax and head, while the legs are brown. This beetle is about half an inch long. The eggs are laid in the spring, and again the latter part of the summer. The larvæ hatch soon after the eggs are laid and develop into short thick grubs not unlike the larvæ of the potato bug in shape, as indeed they may be, for the beetle is allied to that insect. These grubs are light above with a row of black dots along each side, and a black line along the back. Both beetles and larvæ feed upon the leaves and sometimes do considerable damage. The usual remedies would be sufficient. They feed upon the basswood, willow alder and elm. A small snout beetle, *Magdalis armicollis*, Say. is said to infest the elm, but I do not know anything of its habits.

Another beetle known as the

RED SHOULDERED SINOXYLON,

though attacking fruit and other trees is also found upon the elm. This beetle is one of a large group of boring beetles. Its scientific name is *Sinoxylon basilare*, Say. It is black with a red spot on

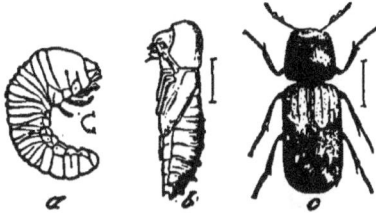

Figure 49.
Sinoxylon basilare, Say.
a. larva, *b.* pupa, *c.* beetle. All enlarged.

the anterior portion of each wing cover. The form of this insect and general appearance is shown in figure 49. The larva is white or yellow. It bores into the wood of the tree, which it infests, and if in any considerable numbers may do considerable harm.

Another boring beetle which confines its labors chiefly to the bark of the peach and other trees, is *Phlœotribus liminaris*, Harr. This is said to attack the elm, and has ever been called elm-bark beetle, but it appears doubtful if it ever attacks the elm. Another little boring beetle, *Hylesinus opaculus* L. C., very closely re-

Figure 50.
Hylesinus opaculus, Le. Con.

sembling the preceding, is shown in figure 50, taken from Riley. The color is black, and the wing-covers are dotted with little pits. It bores in the bark of the elm and ash.

Dularius brevilineus, Say.

This is one of the largest of the elm infesting insects. The perfect beetle, figure 50, is nearly an inch long, with large, stout legs. The wing-covers are relatively smaller than in most beetles, and do not cover the end of the abdomen. The color is dark blue. Its larva bores into dead wood, and cannot, therefore, be called injurious, unless it should be destructive to timber.

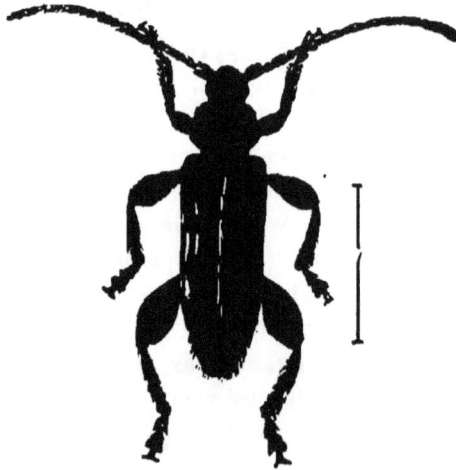

Figure 51.
Dularius brevilineus. Enlarged.

Of all the beetles the most mischievous to the elm, and one of its worst foes in this region, is the

ELM BORER.

This beetle, shown with its larva in figure 52, is somewhat like the common apple tree borer in its general form and appearance, though it is rather smaller, and differently marked. It may be recognized without difficulty by the peculiar dull red-line which

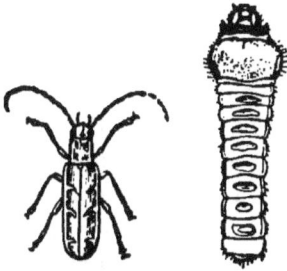

Figure 52.
ELM BORER.
Saperda tridentata, Oliv.

borders the outer edge of the wing-covers, and sends inward, as the figure shows, three curved, tooth-shaped projections. On account of these it might be called the three-toothed borer, if it were understood that the teeth were merely color-marks. Its scientific name is given it because of these three points of color, and is *Saperda tridentata*, Oliv. The general color is dark brown, and besides the red lines and points mentioned, there are other red markings on the thorax. There are also black dots on

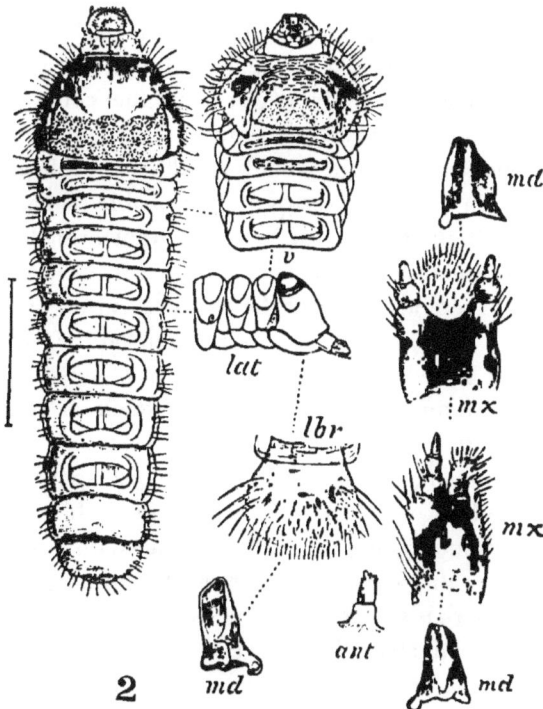

Figure 53.
LARVA OF ELM BORER.
Saperda tridentata, Oliv.

Much enlarged. *v.* under side of head, and five following joints; *lat.* side view of the same; *md.* jaws, or mandibles; *mx.* maxillæ; *ant.* antenna; *lbr.* labrum.

the thorax and wings. A fine grayish down covers the body, so it appears lighter than it otherwise would. As in most boring beetles, the antennæ, or feelers, are long and slender. It is not the beetle, however, but its larva that does the mischief. This is seen at the right of the beetle in figure 52, and an enlarged figure of the larvæ, both from Dr. Packard, in figure 53.

This latter figure gives a very excellent idea of the structure of this borer. The actual length of the worm is about three-fourths of an inch. As may be seen in the figures, its form is flattened cylindrical, the back being less convex than the sides. The beetles lay their eggs on the bark of the trees, and as these hatch the young larvæ bore inward, but they do not seem to ever penetrate as deeply as do the apple tree borers, or many others. They do much of their burrowing in the outer sap wood. During the summer they feed continually, but as cold weather approaches they become dormant, in which state they remain through the winter, again arousing themselves to activity as warm weather comes on. Thus they live, growing larger each season, until the spring of the third year, when they cease boring finally, and become beetles, which appear usually in June. It seems quite probable that all the larvæ do not wait until the third year, but are transformed the second spring. Trees infested by this beetle, which is only too common in Vermont, if there are many of the pest present, at once begin to show signs of trouble, and if the evil has gone on the bark becomes dry, and can be pulled off in large pieces, as it is separated from the underlying wood, and if unchecked, the attack in time not only weakens but destroys the tree. Early in the course of the beetles their presence can be recognized by the moist spots on the bark where the sap has oozed from the borings. If such places be taken at once in hand, and the bark cut into with a sharp chisel, the borers can be discovered, and removed with no great injury to the tree. If the bark is scraped and kept as smooth as possible, which in the elm, unfortunately, is not very smooth, and then washed over with kerosene or carbolic emulsion early in June, or perhaps the last of May would be as well, the female beetles would not be likely to deposit their eggs upon it. Cutting out the larvæ is a tedious process, but when they are once lodged in the wood it is about all that can be done, and a badly infested tree can only be treated successfuly as it is cut down and burned, so that the beetles may not spread from it to other trees. The downy and hairy woodpeckers are undoubtedly of especial service in destroying these borers, and the tit-mice as they hop over the bark probably destroy their eggs. Another allied species, *Saperda lateralis*, Fabr., infests the elm, but as it confines its attention to dead trees, it does no harm.

Another beetle, *Neoclytus erythrocephalus*, Fabr., reported as boring elms in Michigan.

A leaf eating beetle which thus far has confined its devastations to the slippery elm is,

Monocesta coryli, Say.

Figure 51, after Riley, gives a very excellent representation of this insect in its various stages. The beetle is light yellow with " two dark bluish spots on each wing cover." These spots, however, are sometimes entirely lacking. The larvæ are of the form shown in figure 51 *d*., and when fully grown are brown, but when first hatched they are yellow. In August or the last of July the larvæ go to the ground and form below the surface a cell, in which they pass the winter dormant. The following spring they change to pupæ, figure 51 *i*., and in about a week they become perfect beetles, which appear in June. Spraying infested trees in June, or as soon as the eggs begin to hatch, would be efficient, and Dr. Riley says that as the larvæ are very sluggish they may be shaken from small trees.

There is a large larva which has attacked elms in this neighborhood, but I cannot say what it is as I have not been able thus far to capture it. My attention was first called to its ravages last fall and one of the larvæ, the only one apparently in the region, had been taken, but was destroyed before I returned from out of town. The description given by several who saw it would indicate the larva of some beetle like *Prionus*, and I am inclined to think that it was *Prionus brevicornis*, Fabr. of a closely allied species. Its method of attack, however, was different from any that I have seen, judging from its results. There were around the tree from which the larva that was secured was taken, furrows in the inner bark and outer wood, some of them two or three feet long, and callused over by the growth of the bark an inch or even two inches wide. Of course the original furrow was less than this, but the nearly fresh one which I saw was three quarters of an inch wide. Evidently the tree had long been the home of this

Figure 53.

Monocesta coryli.

a. Mass of eggs; *b.* egg magnified; *c.* larva feeding; *d.* adult larva; *e.* larva molting; *f.* outline of one joint; *g.* head of larva, enlarged; *i.* pupa; *j.* beetle.

species, for there are about it great callous ridges from near the ground to up on the larger branches.

ORTHOPTERA.

This group, although it contains many insects that are exceedingly destructive to crops, contains few that injure trees, and I know of only one that infests the elm, and this is nowhere, I think, very troublesome. This is the,

TREE CRICKET.

It is not a very common insect here, though perhaps more so further south. Figure 55 gives a view of the male of this species. *Œcanthus niveus. Serv.* It is more destructive to raspberries and grapevines. The males are white and the females sometimes white, sometimes dark. The insect is allied to the common cricket, and is able to produce very loud, shrill chirping. The female pierces with her ovipositor the canes of raspberry or wood of small branches, peach or grape or elm, and deposits her eggs in the peth. These punctures are in a row and may so weaken the twigs that they break off. The insect does not eat any part of the plants which it infests and the injury which it does is purely mechanical. They are also to be ranked with beneficial insects, for the larvæ feed upon plant lice and other small insects. Whether they on the whole do more harm than good I do not know.

Figure 55.
Tree cricket.

HEMIPTERA.

To this group, which embraces some of the most simply organized insects and the lowest in rank, belong some very destructive species. All the insects of this group which we shall consider are small, most of them microscopic, and their life history is very complex and obscure. Those which infest the elm are such as are commonly known as plant lice. No group of insects contains so many species which are likely to be overlooked because of their minute size and because they hide in crevices of bark and similar locations, and this is the more unfortunate because, when once established, they multiply prodigiously and soon overrun a tree to such an extent that their removal involves much labor. The amount of injury which any single plant louse, or aphis can inflict is quite trivial, but when this is multiplied by the large numbers which may exist, the evil reaches large proportions. Sometimes the aphides can be eradicated without great difficulty, sometimes they defy all efforts with any known insecticide. They all have a beak with which they puncture the plant upon which they have settled and having thus made an outlet for the sap they feed upon it. This drain upon the plant weakens it, and when the leaves are attacked they may swell, crumple or become covered with galls so that their natural function is hindered or prevented, and the plant suffers if it does not perish. The aphides are eminently plant parasites, and almost all species are infested by one or more species of aphis, which are often peculiar to it.

The " honey dew" found, it may be to a disagreeable extent, about plants infested with aphides, comes from the repletion of the insects. They suck up the sap of the plant on which they have located until they are full and more. In many species there are on the upper, hind parts of the abdomen two little tubes which serve as outlets for the sap, and this is distributed as a transparent, milky or colored liquid of a sweet, sticky nature, much liked by ants and some other insects. .This " honey dew" is sometimes produced to an astonishing extent, considering the very small size of the insects which produce it. Last August it was literally showered down from some of the elms in Burlington, so that the fences, walks or whatever was beneath them were bespattered thickly with the sticky fluid which turning black as it was exposed to the sun and air and became exceedingly unpleasant. As will be shown later the metamorphoses of the aphides and their allies are much more numerous than is usual, even among insects, and the differences are sometimes so great that it is difficult to recognize the same insect under its varied disguises. Not only are there different broods, each different from the others, but within the limits of each brood the changes from the egg to the adult are often great. So, too, in their reproduction they vary. One brood lays eggs, which perhaps develop into forms unlike the parent, all of which produce living young, there being no sexual distinctions, and in course of time from these, or their descendants, come true males and females which pair and again hermaphrodite forms appear for a series of generations, and so on. Fortunately for man the aphides are weak and slow moving insects, and are easily seized by their numerous enemies.

The common laced-winged fly, an insect often coming into our houses on warm evenings, shown with its eggs in figure 56, is a

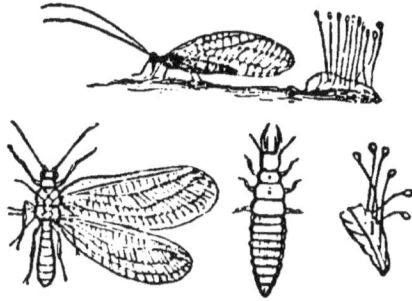

Figure 56
Lace-wing flies, eggs and larvæ.

very useful insect as it feeds on plant lice. The insect is a very pretty, gauzy-winged creature of a delicate green shade, but its attractions disappear if the contact with it is too close, for it produces a most disagreeable odor. Dr. Packard says, that this insect, *Chrysopa*, is so much esteemed as a destroyer of plant lice, that European gardeners " search for these aphis lions and place them on fruit trees overrun with lice, which they soon depopulate." There are a number of species of " lady-birds " or " rose-bugs," small reddish globular beetles with black spots and markings, though some are differently marked. Figure 57 gives several dif-

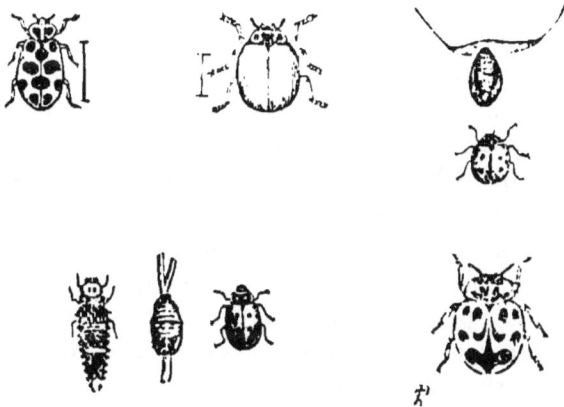

Figure 57

ferent species of the lady-birds and shows differences in the markings as well as in form. The first figure beginning on the left in figure 57 is *Hippodamia maculata*, De. G. a common species. The next without spots is *Coccinella munda*, Say. and the next *Coccinella novemnotata*, Herb., above which is its pupa. Below, on the left, is shown a form common south of here in three stages. Another figure of the same species is shown in figure 58 number 14, natural size and enlarged. Figure 58, copied from a plate by Prof. Comstock, gives excellent illustrations of several species of these lady-birds both natural size and enlarged, and the larvæ of most of them, enlarged. A brief enumeration of the species illustrated may be of service to some. As I think it very important that all should be familiar with these insects, I have introduced numerous illustrations and have thought best to give figure 58 entire, as Prof. Comstock gave it, though all of the species figured are not found in Vermont. Numbers 1. 2, 3, *Cycloneda abdominalis*, Say., larva, pupa, and perfect beetle, "is a small ashy gray insect, there are seven black spots on the thorax and eight upon each wing cover." Comstock. *Cycloneda sanguinea*, L. is shown in the larva and perfect stage in 4 and 5. "Its color varies from blood-red to brick-red, thorax black with two orange spots and edged with the same color, and head black with two light spots." This is the same species as that shown in figure 56, upper middle figure, named *Coccinella munda*, Say., but referred by Prof. Comstock to Linneus' species. It is common everywhere. Number six is, *Cycloneda occulata* a species which closely resembles a more common one known as the twice-stabbed lady-bird, *Chilocorus bivulnerus*, Mul., as does also number 9 which is a west coast species, shown in its three stages. It lives on cactus. The twice-stabbed lady-bird is especially useful in destroying bark lice. Numbers 10, 11, 12 illustrate a very common and very useful lady-bird, *Hippodamia ambigua*, Lec. The larva of this number 10 is a slug-like insect of a blackish color above, and greenish below. The upper part is spotted with orange. The perfect insect, number 12, is black with a partial white border on the thorax, while the wing-covers are red. As the figure shows, this species is longer than many lady-birds. Number 13 is a western species. Number 14, as has been noticed above, is *Hippodamia convergens*, Guer. This is one of the most common species, especially in the south. The only beetles which

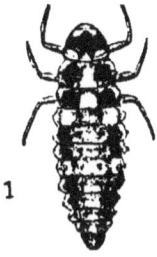

1

2

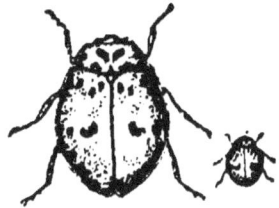

3

4

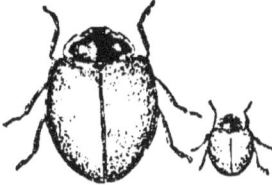

5

6

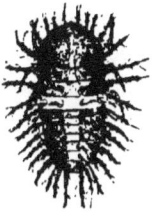

7

8

9

10

11

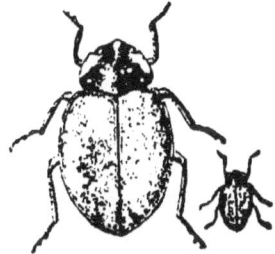

12

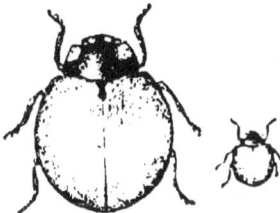

13

14

A.B. Comstock del.

Figuro 58.

I now recall that are at all likely to be mistaken for the useful lady-birds are shown in figure 59. The left hand beetle is one of

Figure 59.

Epilachna borealis, Th.   *Diabrotica 12-punctata,* Fab.

the lady birds, but it does not, as all the rest, feed upon insects, being a vegetable feeder. Its name is *Epilachna borealis,* Th. It is, so far as I know, the only one of the many lady-birds which feeds upon vegetables. It is larger than the other species given, of a reddish yellow color with seven black spots on each wing-cover. The head and thorax are colored like the wing-covers. It feeds upon nearly the same plants as do the common striped cucumber beetles. Another beetle resembling rather closely the lady-birds is *Diabrotica 12-punctata,* Fab., which eats cucumber, squash and other leaves. This is shown at the right in figure 59. A group of flies known as Syrphus flies, one of which with its larva, both enlarged, is shown in figure 60, are useful in destroying plant

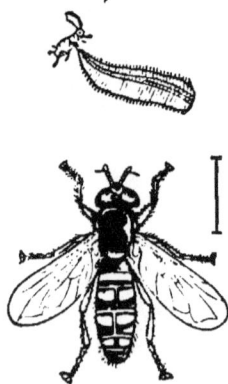

Figure 60.
Syrphus fly.

lice and many of the carnivorous insects find acceptable food in these soft bodied insects.

Among birds, perhaps the tit-mice are as serviceable as any in destroying the plant lice. Insecticides of various sorts have proved more or less useful preventives of the ravages of plant lice. Pyrethrum, especially if used mixed with water, is often very good. Potato parings boiled in water produce a liquid which appears to be fatal to many insects. The kerosene or carbolic acid emulsions mentioned in the early pages of this article are very useful against many of the plant lice. Some years ago a solution was recommended by the United States Agricultural Department which is 'made as follows :—

> 2 ounces Sulphur flour
> 2   "   Washing soda
> ½   "   tobacco

A piece of quicklime the size of a duck's egg. Pour all the ingredients into a pot or pan, boil for fifteen minutes stirring constantly. After this let the mixture stand quietly until it settles, when the clear upper portion can be poured off from the sediment. The portion poured off will keep indefinitely. In using it needs to be diluted considerably, more or less as the foliage is tough or tender.

The plant lice are grouped by entomologists in four subdivisions, or families. The first family includes the familiar scale insects, bark lice, mealy bug, etc., and is called coccidæ. It is a group not easily defined because it includes many varied forms. In many cases there is a remarkable diversity between the males and females and sometimes they are to all appearance about as unlike as possible, certainly they are more unlike than many insects placed in widely different zoological groups. This family includes a large number of exceedingly troublesome insects, and no insects are more readily carried from place to place since in some of their stages of growth they are able to withstand great changes in the surrounding conditions. Moreover some are protected by a scale which completely covers them, others secrete a downy or wooly covering which wholly changes their appearance and also affords some degree of protection. Certain products of commercial value as cochineal, and shellac are obtained from species of coccidæ and quite a number secrete wax sometimes in sufficient quantities to be of value. As a rule the perfect males are active, two-winged insects, while the females are sluggish and wingless. To this group belongs the oyster shell back louse of the apple and other trees which has long been known as *Aspidiotus conchiformis*, but Dr. Riley has removed it from this species and redescribed it as *Mytillaspis pomicorticis*, Riley, while he retains the old specific name for the bark louse of the elm which thus remains *Mytilaspis conchiformis*, Gmel. It however, certainly resembles very closely the apple scale both in habits and appearance.

THE ELM BARK LOUSE

has long been known in Europe, and was doubtless brought to this country on imported elms, and has not, at least so far as I know, increased to such an extent as to be seriously injurious to the American elm, on which, however, it sometimes occurs in considerable numbers. In this species both males and females are covered by a scale soon after they are hatched, though the females far outnumber the males. The female scales are familiar to all, and in this species the male scales are not very different. If in the winter or early spring the scales are examined, they will often be found to cover what appears like a yellow or reddish dust, but which is really a mass of eggs. These eggs hatch in the spring or early summer, into minute, active larvæ, which soon leave the protecting scale and run over the tree. At this time their form is oval, and they may be not more than $\frac{1}{100}$ inch long, so that they are only just discernible by the unaided eye. They have the usual beak of this class of insects, and this they thrust into the bark, and begin to pump up the sap. The insect is soon attached by its beak, and after this does not move about. Ere long, delicate white threads are produced, and later, a continuous covering which becomes the scale. The form changes, and the insect is soon very different from what it was when first hatched, and on the whole, it appears to be less highly organized, if the insect be a female. After the impregnation the eggs develop in the body in great number, and these, when mature, are deposited under the scale. The male develops differently, and is, apparently at least, more highly organized than the female. He is furnished during a part of his life with a gauzy, transparent pair of wings, the hind wings being developed only into little hooks. After laying the eggs the female perishes, and during the winter only eggs are found in the scale. The perfect males have no proboscis or mouth parts, as the female has, but by a most singular transformation the mouth parts are

replaced by a pair of eye-like bodies, which probably serve as a
secondary pair of eyes. Prevention is eminently better than cure,
in case of scale insects. If they are only detected when they first
appear, they can be readily removed by a blunt-edged knife or
stick. If too late for this, washing the twigs with strong lye water,
kerosene emulsion, or some such substance, is beneficial. Very
likely other species of scale will be found on the elm as the group
is more carefully studied. A new insect not known in this country
until within a few years, is allied to the scale insect, though instead
of a scale it protects itself by a cushion of cottony fibres. This
may properly be called, though the name is not very distinctive.

IMPORTED ELM LEAF APHIS.

This was first described in this country by Mr. L. O. Howard,
Assistant Entomologist United States Agricultural Department,
and to him I am indebted for the accompanying figures. It was
first discovered in 1884, in Westchester County, N. Y., and from
there specimens were sent to Washington, and afterward other
specimens from other localities. Mr. Howard, far more completely
than any one else, has worked out the life history of this insect,
and to his account published in *Insect Life*, vol. 2, p. 34, we are
much indebted. The insect has been gradually extending, and
now has quite a wide range over the country. How long it has
been in Vermont I do not know. I first found it in Burlington in
1889. Whether our specimens have come from other infested
American elms originally, or from European elms, which have
been planted here, of which there are a few, I cannot tell, but it is
here at any rate, and its history is of sufficient interest to demand
some attention. It is reported as abundant near Boston. Its
food habit seems to vary somewhat, for we are told that in Cambridge, Mass., it is found upon the slippery elm : in Washington
it occurs upon the European and several other species. Here it
occurs on the common American elm, and I have followed it for
some time in its development upon this tree. During the late fall
and winter we find it dormant in the crevices of the bark. Specimens which I examined in November, or in some cases earlier,
and others collected in January and March, were precisely alike,
having the form and appearance of those shown in figure 61.

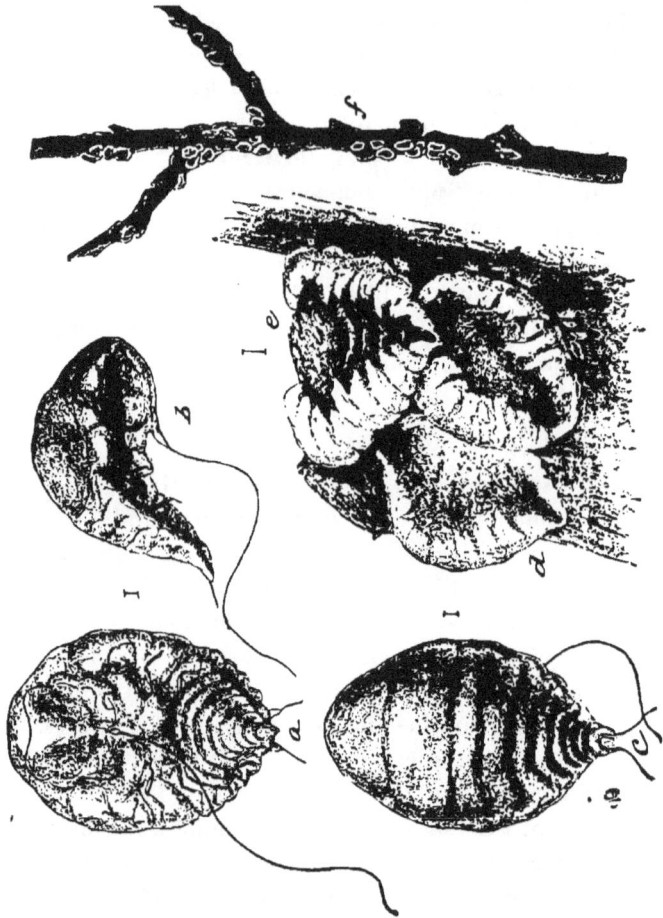

Figure 61.

Gossyparia ulmi, Geoff.

a. b. c., appearance of the insect as it is dormant during the winter; *e.* cushions, or scales on the bark upon which the insects rest; *f.* twig, with insects on the bark.

At this time those which I have seen are of a light, yellow color, and have a sort of resinous appearance,—that is they seem as if made of gum copal or amber. They are usually, to all appearance, lifeless, although when kept for some time in a warm room, they curl as shown in the figure at *b*. It is hardly necessary to notice that the figures are all greatly magnified, the lines by the side of each figure being intended to indicate the actual length.

In the spring as the warm weather comes on the dormant lice awaken to activity and as Mr. Howard informs us the females then

throw off the old skin and the males form a cocoon of waxy filaments which they secrete. This cocoon is an oblong cylindrical
affair, six or seven hundreths of an inch long, in which the insect
remains for some time. There seems to be two sorts of males, one

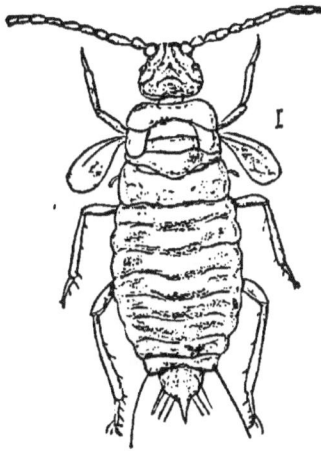

Figure 62.

Gossyparia ulmi.

Imperfect male.

of which figure 62, comes from
the cocoon earlier than the other
form. As the figure shows, it has
very slightly developed wings, so
small that they can be of no use
to the insect. This form seems to
be sexually perfect; although this
point is not certainly established.
They certainly pair with the females, and probably impregnate
them. The body of these males
are about six hundreths of an inch
long and fifteen thousanths of an
inch broad. The true males figure 63, middle figure, are furnished with a pair of large wings
by which they fly readily. As has been indicated above, they
are longer in the process of pupation than the less fully winged
males coming from the cocoons some days later than they do.
The different form of the body, its more slender shape, and especially the two long filaments extending from the abdomen of the
perfect males will be readily seen in the figure. It does not appear
clear whether the two classes of males develop from different
sorts of eggs or whether the perfect males are those in which additional development and moulting has occurred. Mr. Howard
evidently thinks the latter to be the true explanation of the case.
Judging from my own observations I think it probable that some
differences in the food, season, climate, etc., affect the development of the insects and that under some conditions imperfect males
predominate, while under others the perfect ones are more abundant. In Europe it would appear that either the perfect males are
very scarce or have been overlooked by entomologists, if we may
judge from their writings.

As figure 63 a. shows, the female before pairing is very unlike
the other forms, and her own changes greatly after impregnation.
She is at first a spiny oval, wingless insect the segments of the

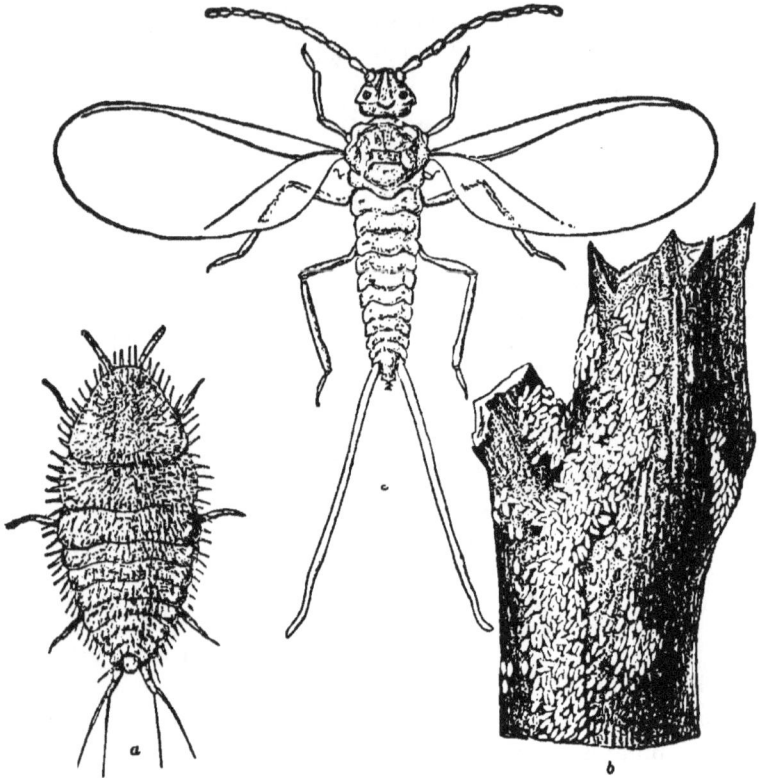

Figure 63.

Gossyparia ulmi.

a, Female before impregnation; *b*, bark covered with male cocoons; *c*, perfect male.

body being very clearly marked. Each joint is provided with glands, which secrete a waxy substance. After pairing, the body becomes more rounded, the waxy secretion increases, the insect remains attached to a given spot, and a thick, scaly cushion is formed, the spines disappear and the surface of the body is smooth and shining. At this stage honey dew is formed in very considerable quantities from the excess of sap which the insect takes from the tree. This is showered down upon whatever may be below the branches on which the insects have located. After a short time the young are produced alive, in June or July.

After the young appear the female loosens her hold upon the twigs or bark. The young are as unlike either parents as these are unlike each other, though they resemble the female rather than

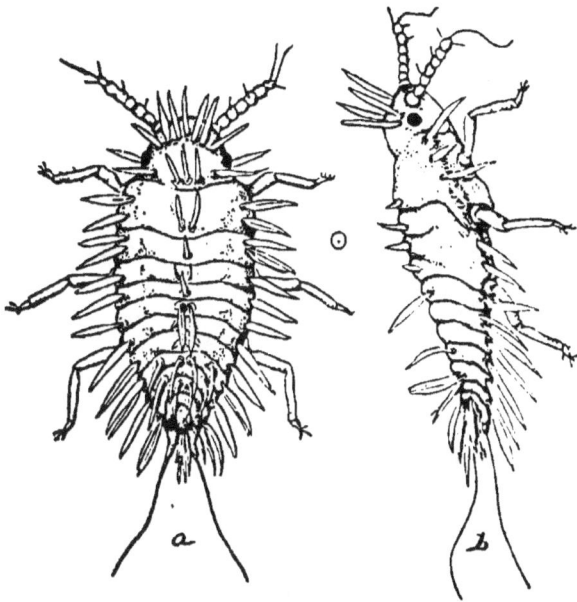

Figure 64.
Larva of Gossyparia ulmi.

the male. Figure 64, gives an enlarged view of the larva seen
from above and from the side. These larvæ are, as might be expected, exceedingly minute objects. They are quite active and
run about the tree looking like grains of fine yellow powder. After
running over the branches for a short time, this in July here, they
settle themselves on the under side of the leaves on young twigs or
about buds. According to Mr. Howard. in species like slippery
elm, in which the twigs are hairy or pubescent, they settle on the
leaves, while in the smoother species they settle on bark or twigs.
In August the young lice leave their summer feeding ground and
go to the large branches or trunk, in the irregularities of which
they find hiding places. At this time they produce again quantities of honey dew. Last August this was very abundant beneath
some large elms in Burlington. According to Mr. Jack, of the
Arnold Arboretum, this honey dew gives out a pungent odor which
can be perceived at some little distance, but I have never been able
to discover it. As cold weather approaches, the dormant stage with
which we commenced is again reached and the insects are in the
condition shown in figure 61. This insect has been found on the

American elm, two species of European elm, the slippery elm and the cork elm. *Ulmus racemosa*.

In regard to the presence and spread of the insect under consideration Mr. Howard says:

" The finding of *Gossyparia ulmi* upon American elms and upon European elms in this country was quite to be expected, and the only wonder is that it has not been found and recognized before. The species of Coccidæ have already extremely wide ranges, and every season still further extends them. Of our admitted North American Coccid fauna twenty-three species are of European origin (one more doubtfully so), three are from Australia and New Zealand, while sixty-nine are either truly North American or their original home is unknown. As several of these are found only on hot-house plants, they are certainly not North American. Several others are found on both native and imported plants and there are no data upon which to decide upon their proper faunal position. The fact that the *Gossyparia* prefers American elms at Cambridge is by no means without precedent in the group, and as another instance it may be mentioned that the beautiful oak-scale *Asterodiaspis quercicola* (Bouche), recognized by Comstock in 1880 upon foreign oaks on the Department of Agriculture grounds, is at the present time to be found almost solely upon American oaks in the same grove." *Insect Life* vol. 11 p. 41.

In considering remedies, perhaps not very much need be added to what has already been given among general insecticides. In the spring and early summer, kerosene emulsion or carbolic acid emulsion sprayed over the trees would prove a check. Mr. Jack says that as Arnold abortum whale oil soap and kerosene were used successfully on the trunk and larger limbs and twigs. The lady-birds already fully noticed should not be overlooked in this connection. On the same trees which were infested with *Gassyparia* I found at different times during the winter, groups of lady-birds also hibernating in the same crevices and doubtless the beetles were awake as early in the spring as were the lice ready to feed upon them and to lay eggs among them which should soon hatch into larva that would be veritable lions and tigers to the young lice. The species most common was *Adalia bipunctata* the two spotted lady-bird and I have no doubt that these beetles are most serviceable in keeping the plant lice in check. In California where the orange scale has been very troublesome, the imported Australian lady-bird has been found very useful so that the State Board of Horticulture raised the beetle for distribution among orange growers, that by its aid the scale might be checked. How valuable an aid this is, I do not know, as reports have not yet come in very extensively.

The second family of plant lice is called in the books the *Aleyrodidæ*. Less appears to be known concerning this family than of the others. They are winged in both sexes. The color is usually light, white or whitish, or yellow, and the distinguishing character is found in the light powder which covers the bodies of the adult insect. In the immature state they are covered with a scale. I do do not know that any species of this family is found on the elm.

The third family are the *Aphididæ*. These are the well-known plant lice which are more familiar to most than others of the group. The common green fly of the gardeners may be taken as a type of this family. A considerable number of injurious insects belong to this group. In his excellent Synopsis of the *Aphididæ* of Minnesota, Mr. Oestlund has enumerated five species of *Aphides* which are found upon the American elm, and one which is confined to the slippery elm. It is more than likely, however, that when the group has been more thoroughly studied, this number will be increased. The family embraces a very large number of species infesting a great variety of plants. All are very small : most are soft-bodied, delicate insects, with four very thin transparent wings, or in many stages without wings. Like the *Coccidæ* these insects secrete honey dew at certain times in considerable quantities, and all are familiar with the fondness which ants manifest for this sweet substance. Other insects also feed upon it, but ants more than others. The changes which the *Aphides* undergo as they pass from brood to brood, will be better understood by the illustrations which follow, than by any general description. As will be seen, their modes of multiplicaton are peculiar. Many of them so affect the parts of plants which they infest that galls, swellings, crinklings, and other deformities are produced. The first species of this group to be considered is

THE ELM LEAF CRUMPLER.

This insect, *Schizoneura americana*, Riley, is of much interest,

18

both economically and scientifically. It has been made the object
of very careful and thorough study by Dr. Riley, and to his paper
we are indebted for the cuts which illustrate the species, as well as
for many facts. This insect attacks chiefly the under side of elm
leaves, and by its punctures causes them to swell and crumple, as
shown in figure 65.

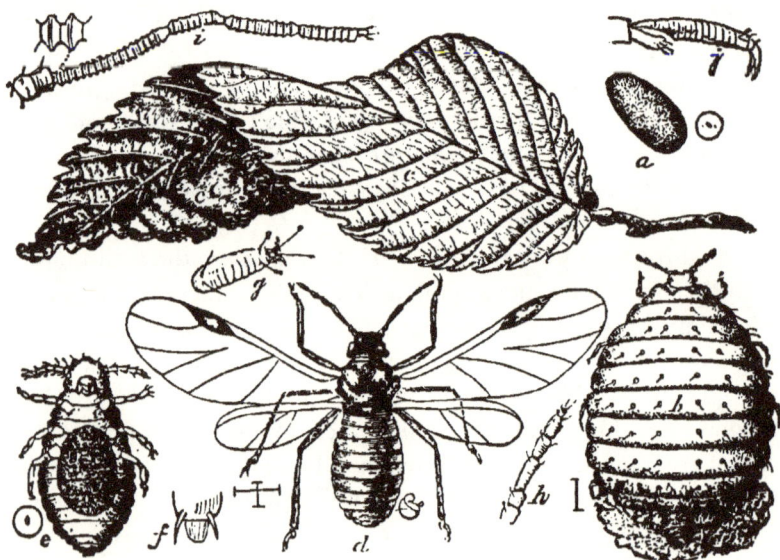

Figure 65.

Schizoneura americana, Riley.

ELM GALL LOUSE.

a, egg, as found in the winter; *b*. female in the spring; *c*. leaf curled, and with galls;
d. winged female; *e*. wingless female; *g*. tarsus of *e*.; *h*. antenna of fourth
generation; *j*. tarsus of *d*.; *i*. antenna of *d*.

The leaves are often very much more crumpled and rolled up
than that shown in the figure. Indeed, I have found this species on
young leaves, which they had so affected that they were almost
globular. The eggs are minute yellow bodies, figure 65, *a*., which
deposited in the crevices of the bark are often covered and partially protected by the dried skin of the mother. The eggs are
about two hundredths of an inch long. In the spring when the
buds begin to swell and open, these eggs hatch, and the young
larvæ at once make their way to the ends of the twigs and attach
themselves by their beaks to the young leaves on the under side,

and the irritation caused by them affects the leaves as we have seen. There is no regularity about the curling; different leaves curl differently, and those which I have seen here, as has been noticed, curl more than is shown in figure 65. The curled leaf affords shelter to the *aphides*, and on opening it multitudes of them may be found hidden within its folds, little grayish or blackish bodies. When first hatched the larvæ are light colored, but as they grow older they grow darker; the legs are always darker than the body. According to Dr. Riley, the individuals of this first generation complete their growth in about twelve days, though the time varies with the season, for if cold days come the insects are dormant for a time, and then development is retarded. There are probably, Dr. Riley thinks, three molts. These, when mature, are the " stem mothers," figure 65, *b*., and they live in the galls which they have produced, and commence filling the galls with young, which are alive when produced, and come at the rate of " one every six or seven hours, according to temperature, increasing in bulk and prolificacy from day to day, until by the early part of May in the latitude of St. Louis, she has attained her fullest development, and soon perishes. Riley, Bullet. United States Geological Survey, Hayden, v, p. 5.

The young of the second generation, those which come from the offspring of the winter eggs, the " stem mothers " of Riley, are very much like the parent, except that they never attain to so large a size, and are rather different in color, being of a reddish brown, or liver color. From these the third generation comes, and the individuals of this generation differ considerably from the preceding. These latter are wingless, and are of the general form shown in figure 65, *b*., while the third generation are like *d*. of the same figure, and, as the figure shows, almost all the proportions are changed. The wings do not appear at first, but are developed as the insect becomes mature. Of these early generations Riley says : " During most of the month of May we may find, where large clusters of leaves are affected, the few more or less exhausted stem-mothers, and these second and third generations in every stage of development. As the lice increase in number the leaves no longer protect them, but present on both sides multitudes of busy atoms, livid old, and paler young, those with wings and those getting wings, interspersed with white exuviæ, cottony secretion, and globules of pearly liquid. At the same time in single curls of more terminal leaves, we may find the second generation of wingless mothers, surrounded by smaller colonies, all of which will become winged." These winged females of the third generation do not appear to be long-lived, but during their life they produce what may be regarded as living young in great number. I say, what may be so regarded, for although deposited as eggs, and having the appearance of eggs, yet they hatch into living young almost immediately after they come from the parent insect. The larvæ of this fourth generation seem to be more brisk and active in their movements than any that precede them, and may be seen creeping over the twigs. At first they are reddish in color, but like all the broods they grow darker and browner with age. Like the second generation they have no wings, and like them they attack the leaves, increasing the deformity begun earlier in the season. Nor do they invariably attack the leaves, but appear to thrive on the younger bark, where sometimes they may be found, of course with no sheltering cover such as the gall affords those on the leaves. They are, however, more or less covered by a cottony substance which they secrete, although this in no way takes the place of the gall, since it is far more delicate, and is easily rubbed off. Reproduction goes forward rapidly, and they extend themselves over the twigs with corresponding rapidity. A phenomenon noticed by Riley, has not come under my notice, but it is extremely interesting. He says: " At this season of the year, when the lice are thus numerous, they may be found during the heat of the day actively crawling over all portions of the tree—a veritable migration necessitated by the want of sufficient succulent leaves, but destined to be the death of the individuals participating in it, excessive multiplication here, as in all other cases, obliging the destruction of the excess." From the fourth generation comes a fifth, which does not differ much from it. This fifth generation gives rise to a second winged brood like the third. This stage in the life of the species is reached in July, or sometimes about the last week in June. I have found them the 26th. They collect in crevices of the bark more than on the leaves or twigs, and in these places the seventh and last generation is produced. These for the first time show clearly sexual distinctions, reproduction up to this time having been asexual. They are dark yellow, or orange, and remain upon or in the crevices of the bark. The mouth is less perfectly developed than in former generations, and the eyes are simpler. The general form of a female of this generation is seen at *e.*, figure 65. After a short time the bright yellow, which at first characterizes the insect, becomes paler, the skin meanwhile being cast off, and though at first smooth and glossy, the surface of the body becomes dusty or powdery. The females, which are larger than the males, are about fifteen hundredths'of an inch long. As may be seen in figure 65, *e.* the female as she grows has a single egg, developed within the abdomen; its outlines can be indistinctly made out. Soon after pairing both sexes perish, leaving in the crevices or under scales of the bark the eggs more or less covered by the dried and shriveled skin of the parent, and thus they remain through the winter to develop in the spring into new "stem-mothers."

Schizoneura ulmi, Linn.

Is a very similar species, and infests European elms. This species has undoubtedly been confounded with the preceding, but it differs in some important respects, such as that the stem-mother of the last named species settles on the upper instead of the under side of the leaf. The galls are different, the wings seem to be longer, and the veining is somewhat different. Still the two species certainly approach each other very closely.

Schizoneura Rileyi, Thomas.

Is a species common in some sections of the country, and noticeably different from the other species that attack the elm, and in the character of the injury which it inflicts. It is described under the name of *Eriosoma ulmi*, by Dr. Riley in his First Report on Insects of Missouri; but Dr. Thomas in his monograph of the family, in Vol. 16, Transactions Department of Agriculture, Illinois, states that Riley's name is preoccupied by one of Linneus. The winged form is dark blue, about a tenth of an inch long, with long transparent wings. The younger, wingless forms are lighter, and sometimes are of a reddish shade. From the cottony or wooly substance which these insects secrete Riley calls the species THE WOOLY ELM TREE LOUSE. They collect on the tree twigs, branches, or trunk in clusters, and by the visitation of their punctures, produce a gnarled growth. The wool secreted is of the clearest white, and wholly covers the body of the insect, and when many are located on a branch they give it the appearance of being covered with cotton wool. This makes them more conspicuous than most insects of this group. They cause great damage to the trees at times. Kerosene or other emulsions washed over the tree would be useful. Dr. Riley says, " I have experimentally found that a washing with a weak solution of cresylic acid soap will kill them instantly." Lace winged flies and doubtless other insects eat them freely. Another species which is quite common is one which produces very conspicuous and easily recognized galls on the upper side of the leaves; this is the

COCKSCOMB GALL LOUSE.

Colopha ulmicola, Fitch.

The general condition of leaves infested by this insect is shown in figure 66 *a*. though, of course, there may be more or less of

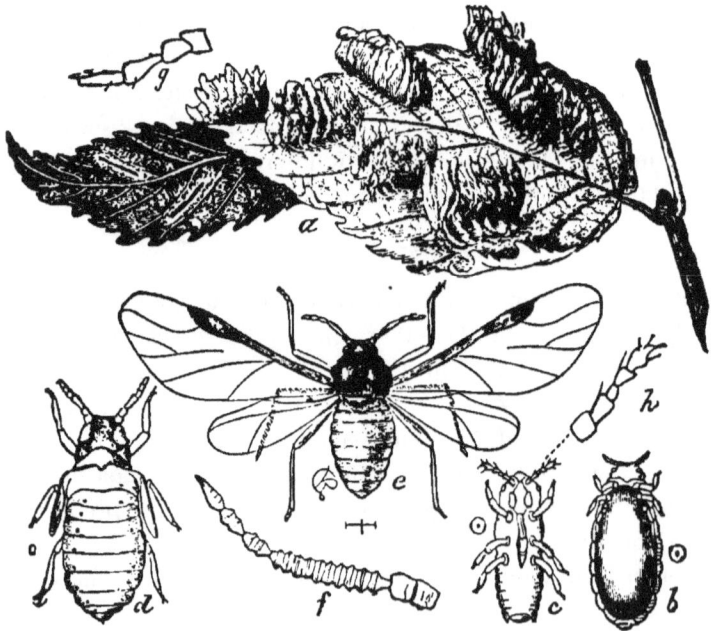

Figure 66.
COCKSCOMB GALL LOUSE.
Colopha ulmicola.

a. Leaf with galls; *b*. egg with dried skin of parent; *c*. young louse; *d*. pupa of *c*.; *e*. perfect, winged insect of *c*.; *f*. antenna of *e*.; *g*. antenna of first brood; *h*. antenna of *c*. All enlarged.

them than on the special leaf figured. They grow with astonishing rapidity, so that one day they are scarcely noticeable, while in a few days they are large. They are usually more or less colored red or brown, and I think the insects, as do most of the plant lice, prefer young trees to old ones. This species was first studied and described by Dr. Fitch in his Fifth Report, as New York Entomologist, but Dr. Fitch's account is by no means complete and has since been supplemented notably by Dr. Riley, whose figure is reproduced in 66. The egg, figure 66 *b.*, is placed in the crevices of the bark, and like that of the Schizoneura, is more or less protected by the dry skin of the insect which produces it. It is a very minute, smooth, yellowish brown body, and remains through the winter, hatching in the spring into a " stem mother," a dark greenish, or olive object which creeps over the twig until it reaches the twigs where the unfolding leaves attract it. To the lower side of these leaves the little insects attach themselves, thrusting their beaks into the soft pulp, and a gall soon begins to form. On the under side of the leaf the opening is a mere line or slit, hardly as conspicuous in many cases as in the figure, but the swelling of the upper surface is very obvious, being half an inch or more high and twice as long. At first the galls are mere linear elevations, the cockscomb-like form does not appear till later. In the early stages of the growth of the gall the originator of the summer's brood, the louse which hatches from the winter egg inhabits it as a domicile, but ere long, as the gall changes the occupant also changes, molting, growing lighter in color, and secreting finally a whitish chaff. As the gall reaches its full size the insect produces young, and this continues until the gall is full and sometimes crowded, and on pulling apart the sides of the slit on the under side of the leaf, they may be seen, little mouse colored or light gray objects of the form of *c.* figure 66. This is about the last of June in this region. Honey dew is produced at this time, though in less abundance than by some other species. In time the lice become of the form shown at *d.* and finally, when fully mature like, *c.* figure 66. In the young of this generation the color is light greenish or yellowish green, but by the time the winged state is reached the color has become dark olive. Until the winged, perfect stage is reached the gall is the sole habitation of the insects, but when thus mature they leave it, and so far as is yet known, all of this generation are females and produce eggs. The galls here are mostly empty about the second week in July, though some of them will be found full later than this and the insects do not all leave the galls at the same time, for I have at the same time found galls in which a few winged insects remained, others in which there were more and others that were emptied. After the galls have been at least partially empty the insects may be found on the twigs and at the ends of the branches, though how permanently they locate there I do not know. There is a gap yet to be filled in the history of the insect between the third generation and the mouthless sexual individuals, the females of which so often perish while containing the single winter eggs. Respecting this, Dr. Riley remarks, " I have not been able to prove absolutely that there are two broods of the gall-making female, and my observations all tend to the conclusion that no galls are formed, except by the stem-mother that hatches from the impregnated egg. I have never succeeded in obtaining galls either by inclosing the winged females in muslin bags tied on the living trees or by similarly enclosing her immediate progeny, though I have succeeded in obtaining without any difficulty an abundance of galls by so inclosing the stem-mother. Moreover, all such succulent galls as this one are produced on the tender leaves only, and I have failed to find them on any but those which develop early in the season. It is true that we find the galls quite fresh, and containing larvæ, pupæ, and winged insects as late as the first week in July, and these late galls, as well as the insects within them, are generally more yellowish than those found earlier in the year; but a careful study of the structure of the inmates shows them to be identical with those found in the earlier galls ; and these late galls are from present knowledge to be attributed to the work of late hatching and late developing stem-mothers, rather than to the work of the third generation. I am inclined to think that this third generation will be found to have a different habit ; possibly feeding upon some other part of the tree without forming galls, and producing in time the true sexual individuals, something as in the case of *Schizoneura americana*." Bulletin U. S. Geol. Survey, Hayden, v, p. 12. While I would not assume to present observations upon this species which should be of the same value as those of so exceptionally skilful an investigator as Dr. Riley, I may yet add that my own investigations of this species confirm the opinions expressed in the paragraph quoted, and I do not at all believe that galls are produced by any other than the stem-mother, for I have not been able to get any evidence that they are so and very much that they are not. I am also sure that, as has already been noticed, the insects after leaving the galls go to the extremities of the branches. However, whatever steps there may be which are yet unknown, there are at last produced wingless females, in which develops a large single egg, as has been seen. .

Another species of aphis, which I have not yet found in this region, is described by Dr. Thomas, as a new species under the name *Callipterus ulmicola*, Thomas. It may be confined to the West, but it is very probable that it occurs wherever the elm grows. It is described as "an exceedingly delicate species" of a pale yellow color. Perhaps the character most distinguishing this species from others is found in the wingless form, in which "along the lateral margins of the abdomen, in front and behind the honeytubes are minute tubercles, each giving rise to a hair; these tubercles are quite distinct, and about one to each segment." Thomas, Trans. Depart. Agricult. Ills., 1878, p. 112.

A very small species, *Pemphigus ulmi-fusus*, W., attacks the upper surface of the leaves of the slippery elm. The adult, winged form is only seven hundredths of an inch long. The color is dark. By puncturing the leaves they form spindle-shaped galls, about an inch long.

In addition to what has been said concerning remedies, it may be said, that since the aphides and their allies are not leaf-eating, but leaf-piercing and sucking insects, applications of Paris green and the like, which are efficacious in the case of leaf-eaters, are of little use, since these are applied only to the surface of the leaves which these insects do not eat and cannot penetrate to the juices of the plant upon which they live. Alkaline washes or such as injure the soft bodies of the aphides are of most avail, although the fumigation with tobacco, as florists fumigate their greenhouses, using a tent to cover the trees, could be made useful if trees were not too large as also the use of hydrocyanic acid already mentioned. In no group of insects is it so important that the remedy be applied promptly, for none so soon get beyond the reach of remedies. And when once they have thoroughly occupied a large tree it is almost impossible to wholly dislodge them.

The fourth family of plant lice, the *Psyllidæ*, needs no description here, for, so far as I am aware, none of the insects which it includes infest the elm, at least to any noticeable extent.

As has already been stated, the writer has in this paper attempted a full list of elm-infesting insects, and it is believed that, while doubtless some species now and then found destroying some part of this tree have been omitted, the list is tolerably complete. It has not been possible to present a full account of each of the species enumerated, in some cases for lack of space, in others because the life history is not fully known, but I believe that in most cases enough has been stated to enable those who desire to to identify the insect under consideration. It has been the hope of the writer that such a gathering together of the insect foes of a single tree might prove useful, not only to those who with him love and admire this magnificent species, the glory of our village streets, the chief adornment of our meadows, and who wish in every possible way to preserve their elms in soundness and vigor, that they may attain their fullest beauty, but that it may also be of some value, as a convenient reference, to students of entomology.